# Copper, Molybdenum, and Vanadium in Biological Systems

Contributors:
B. A. Averill LeB. R. Briggs, 4th
N. D. Chasteen T. R. Gilbert K. Kustin
G. C. McLeod K. W. Penfield E. I. Solomon
D. E. Wilcox

With 67 Figures and 13 Tables

Springer-Verlag
Berlin Heidelberg New York 1983

ISBN 3-540-12042-4 Springer-Verlag Berlin Heidelberg New York
ISBN 0-387-12042-4 Springer Verlag New York Heidelberg Berlin

Library of Congress Catalog Card Number 67-11280



Typesetting and printing: Schwetzinger Verlagsdruckerei. Bookbinding: J. Schäffer, Grünstadt.
2152/3140-543210

# Table of Contents

# Active Sites in Copper Proteins An Electronic Structure Overview

**Edward I. Solomon*, Kevin W. Penfield and Dean E. Wilcox**

Department of Chemistry, Massachusetts Institute of Technology, Cambridge, Massachusetts 02139
and
Department of Chemistry, Stanford University, Stanford, California 94305**

A detailed picture of the unique electronic structure of active sites in copper proteins is essential to our understanding of their biological functions. This has been achieved through a combination of complementary spectroscopic techniques, chemical perturbations, and parallel studies on copper complexes which serve as spectral analogues. These studies provide a basis for the generation of a "spectroscopically effective" picture of the active site and allow a correlation of changes in geometric and electronic structure with variation in function. When combined with high-resolution crystallographic information, single crystal spectral studies allow a correlation of electronic features with specific geometric features and provide a great deal of insight into the bonding at the site.



* To whom correspondence should be addressed
** Present address of all authors

Structure and Bonding 53

## List of Abbreviations

| | |
|---|---|
| CD | Circular dichroism |
| $cm^{-1}$ | Wavenumber |
| CT | Charge transfer |
| Cys | Cysteine |
| ENDOR | Electron nuclear double resonance |
| EPR | Electron paramagnetic resonance |
| ESCA | Electron spectroscopy for chemical analysis |
| eV | Electron volt |
| EXAFS | Extended x-ray absorption fine structure |
| GHz | Gigahertz |
| His | Histidine |
| IR | Infrared |
| IT | Intervalent transfer |
| LEFE | Linear electric field effect |
| MCD | Magnetic circular dichroism |
| Met | Methionine |
| NMR | Nuclear magnetic resonance |
| SCF-X$\alpha$-SW | Self-consistent field X$\alpha$ scattered wave |
| SOD | Superoxide dismutase |
| SQUID | Superconducting quantum interference detector |
| T2D | Type 2 depleted laccase |
| TT2D | Treated type 2 depleted laccase |
| UV | Ultraviolet |
| VIS | Visible |

# I. Introduction

A number of important proteins and enzymes contain copper ions at their active sites. These copper protein sites are associated with a variety of vital biological functions, including oxygen transport and activation, electron transfer, iron metabolism, and superoxide dismutation. The protein ligand imposes an unusual geometric and electronic structure at the copper site. As a consequence, copper proteins often exhibit unique and characteristic spectral features compared to small molecule copper complexes. Studying these spectral features provides insight into the correlation between active site electronic structure and biological function.

One of the major goals of studying active sites in copper proteins has therefore been to understand the spectroscopic features associated with the active site. This has led to a classification of three general types of copper protein active sites based on their unique spectral features: Blue copper, normal copper and coupled binuclear copper. An additional class of copper proteins, the multi-copper oxidases, contains a combination of these three types of copper active sites. A reasonably firm understanding of the optical and EPR spectra of a number of copper proteins has now been achieved[1, 2]. This article presents an overview of these electronic spectral features and their relationship to geometric and electronic structure.

## *A. Ground State*

Before the unique spectral features of copper in proteins can be discussed, the geometric and electronic structures of normal copper must be considered. In an octahedral environment the cupric ion, which has nine d electrons, would possess a degenerate $^2E_g$ electronic ground state (Fig. 1, left). A geometric distortion which removes this degeneracy would produce a more stable electronic structure, in accordance with the Jahn-Teller

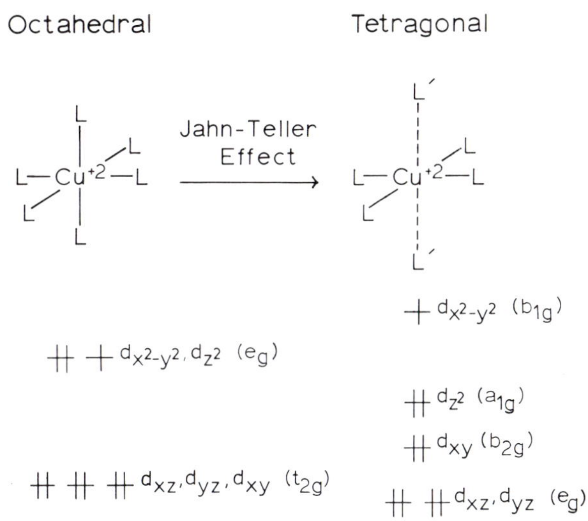


**Fig. 1.** Site geometries and d orbital splittings for copper(II) in octahedral and tetragonal environments (symmetry properties of the orbitals are indicated in parentheses)

Theorem. Consequently, Cu(II) is normally found in a tetragonal environment with four strongly bound equatorial ligands and one or two weakly bound axial ligands. (For a more detailed analysis of $Cu^{+2}$ geometries, see Ref. 35, p. 171–173.) Tetragonal coordination geometry induces a splitting in the d orbitals which is shown to the right in Fig. 1. The $d_{x^2-y^2}$ orbital has the greatest repulsion with the equatorial ligands. Therefore, it is the highest-energy metal-centered orbital, is half occupied, and characterizes the tetragonal Cu(II) ground state as $^2B_{1g}$.

This tetragonal geometry quenches the orbital angular momentum in the ground state by removing its orbital degeneracy. In the limit of a pure $d_{x^2-y^2}$ orbital description for the ground state, this Kramers doublet would be isotropically split in an applied magnetic field, $\Delta E = h\upsilon = g\beta H$, with a spin only g value equal to 2.0023. However, through spin-orbit coupling, the $^2B_{1g}$ is mixed with the $^2B_{2g}(d_{xy})$ and $^2E_g(d_{xz}, d_{yz})$ excited states. Application of a magnetic field to this mixed ground state now gives[3a)]:

$$g_{\parallel} = g_z = 2.0023 + \frac{-8\zeta}{E(xy) - E(x^2 - y^2)} \tag{1a}$$

and

$$g_{\perp} = g_x = g_y = 2.0023 + \frac{-2\zeta}{E(xz, yz) - E(x^2 - y^2)} \tag{1b}$$

where 2.0023 is the g value of the free electron, $\zeta$ is the one-electron spin-orbit coupling constant for copper (equal to 830 $cm^{-1}$ in the free copper ion, but reduced by covalency in the metal complex), and the denominators in the second term of each equation are energy differences between the d orbitals indicated. Note that for a tetragonal copper complex $g_{\parallel}$ governs the ground-state splitting when the magnetic field is oriented along the axial ligand direction and $g_{\perp}$ is obtained when the field is in the equatorial plane. As $g_x$ must equal $g_y$ by symmetry, this complex is described as having an axial g tensor. Frozen glasses of inorganic complexes and protein solutions yield EPR spectra (plotted as the first derivative) which are the sum of transitions of individual Cu centers randomly oriented with respect to the magnetic field. The perpendicular region is more intense due to the greater probability of perpendicular orientation. The net effect is an EPR signal with $g_{\parallel} > g_{\perp} > 2$ as shown at the top of Fig. 2. The quantitative g values observed ($g_{\parallel} \approx 2.2 - 2.3$, $g_{\perp} \approx 2.04 - 2.1$) are in agreement with equation 1 when $\zeta$ is reduced by about 20 percent due to covalency.

The copper nucleus ($^{63}Cu = 69\%$, $^{65}Cu = 31\%$) with a nuclear spin (I) of 3/2 couples to the unpaired electron and produces a hyperfine splitting of the transition into $2I + 1 = 4$ components. This splitting is quite evident in the $g_{\parallel}$ region ($A_{\parallel}$) but is significantly smaller and often unresolved in the $g_{\perp}$ region (Fig. 2, middle). Both isotropic and anisotropic ligand field terms contribute to this hyperfine splitting. The former are Fermi contact in nature and derive from the finite probability of electron spin density at the nucleus. As this cannot occur for an electron in a $d_{x^2-y^2}$ (or any other d) orbital ground state, the isotropic term must derive from an indirect spin polarization of inner filled s orbital electrons (which do have electron density at the nucleus). The anisotropic contribution is a function of dipole-dipole interaction between the nuclear and electron spins and is proportional to $P_d(3\cos^2\theta - 1)$ where $P_d = g\beta g_N\beta_N(1/r^3)_d \approx 350 \times 10^{-4}$ $cm^{-1}$ for

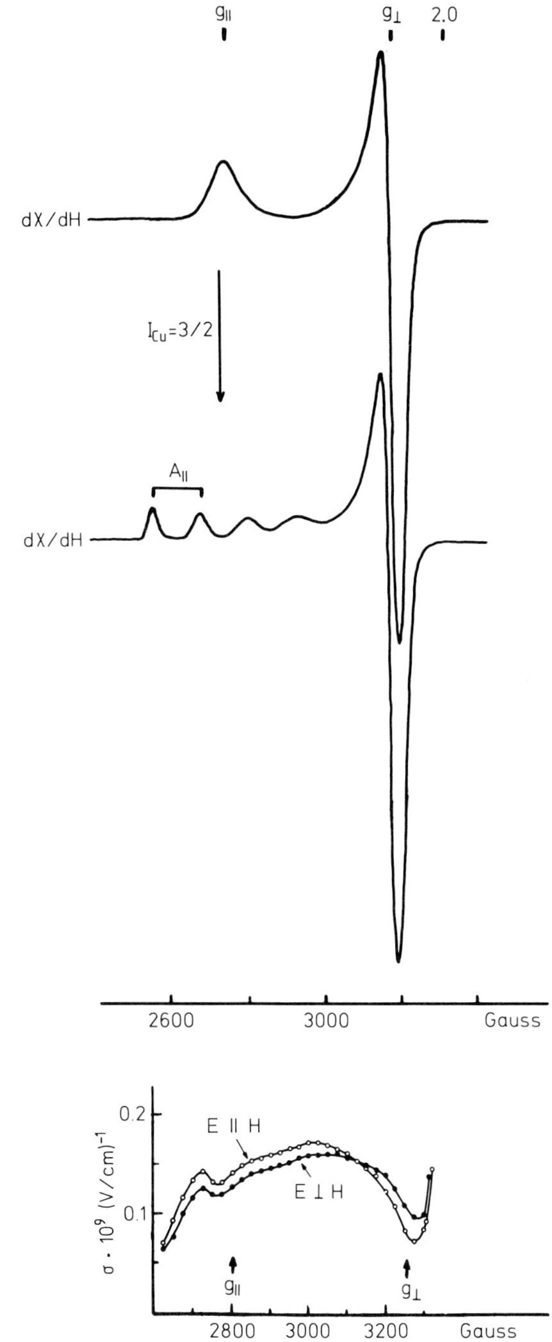


**Fig. 2.** EPR and LEFE spectra for copper(II) in a tetragonal environment. *Upper:* neglecting hyperfine splitting. *Middle:* including hyperfine interaction: $Cu(H_2O)_6^{+2}$ (pH = 2.0; 77 K; $\upsilon$ = 9.25 GHz). *Lower:* associated LEFE spectrum of $Cu(H_2O)_6^{+2}$ (pH = 2.0; 4.2 K; $\upsilon$ = 9.49 GHz) (from Ref. 10)

the free copper ion. $\theta$ is the angle between the axis of the specific ground-state orbital and the nuclear spin vector oriented by the applied magnetic field. These considerations[4] lead to the expressions for the hyperfine coupling given below:

$$A_{\|} = P_d\left[-\varkappa - \frac{4\alpha^2}{7} + (g_{\|} - 2) + \frac{3}{7}(g_{\perp} - 2)\right] \tag{2a}$$

$$A_{\perp} = P_d\left[-\varkappa + \frac{2\alpha^2}{7} + \frac{11}{14}(g_{\perp} - 2)\right] \tag{2b}$$

The first term accounts for the indirect Fermi contact contribution ($\varkappa \approx .35$), the second term indicates the anisotropic dipolar contribution ($\alpha^2 < 1$ adjusts for covalency) and the final term(s) are both isotropic and anisotropic, and result from a coupling of the orbital angular momentum of the electron to the nuclear spin. Substitution of typical tetragonal cupric g values and $\alpha \approx .8$ leads to $A_{\|} = -130$ to $-190 \times 10^{-4}\ cm^{-1}$ and $A_{\perp} = -20$ to $-50 \times 10^{-4}\ cm^{-1}$. The $g_{\|}$ and $A_{\|}$ values are strongly influenced by the ligand environment at a tetragonal cupric site. Representative values for a number of biologically relevant ligands are tabulated in Table 1. As Eq. 1 and 2 indicate, increasing field strength of the equatorial ligands causes $g_{\|}$ to decrease and $A_{\|}$ to increase; increased covalency (often evident with sulfur ligands) leads to a decrease in both values. A more detailed compilation including mixed ligand correlations can be found in Ref. 5a.

Several techniques allow further elucidation of the information contained in an EPR spectrum. Through single-crystal EPR it is possible to determine the orientations of the g and A tensors relative both to each other and to the internal coordinates of a structurally defined active site. The use of several microwave frequencies can be particularly informative. While the spectrum shown in Fig. 2 (middle) was taken with an X-band spectrometer ($v \approx 9$ GHz), Q-band ($v \approx 35$ GHz) and S-band ($v \approx 3$ GHz) should also be employed. The high microwave frequency leads to increased resolution by spreading the g values over a wider magnetic field range with little effect on hyperfine splitting. Low

**Table 1.** EPR parameters for tetragonal copper(II) complexes with biologically relevant ligands

| Ligand Type | Complex | $g_{\|}$ | $g_{\perp}$ | $A_{\|} \times 10^{-4}$ $cm^{-1}$ | $A_{\perp} \times 10^{-4}$ $cm^{-1}$ | Ref. |
|---|---|---|---|---|---|---|
| aquo | $Cu(H_2O)_4^{+2}$ | 2.422 | 2.087 | 134 | <11 | 5b |
| hydroxide | $Cu(OH)_4^{-2}$ | 2.273 | 2.055 | 194 | – | 5c |
| phenoxide | $Cu(Tiron)_2^{-6}$ | 2.280 | 2.066 | 179 | 22 | 5b |
| carboxylate | $Cu(oxalate)_2^{-2}$ | 2.318 | 2.071 | 164 | ≤8 | 5b |
| amine | $Cu(ethylendiamine)_2^{+2}$ | 2.209 | 2.047 | 203 | 28 | 5d |
| imidazole | $Cu(imidazole)_4^{+2}$ | 2.267 | 2.063 | 179 | – | 5e |
| amide | $Cu(biuret)_2^{-1}$ | 2.17 | – | 219 | – | 5a |
| thioether | Cu(1,4,8,11 tetrathio-cyclotetradecane)$^{+2}$ | 2.087 | 2.026 | 172 | 45 | 5f |
| chloride | $CuCl_4^{-2}$ | 2.232 | 2.049 | 164 | 35 | 5g |
| bromide | $CuBr_4^{-2}$ | 2.143 | 2.044 | 190 | 46 | 5g |

microwave frequency often causes hyperfine features to be resolved, as g values, broadened due to site heterogeneity (g strain), become reduced (for example, see Ref. 6). Both Electron Nuclear Double Resonance (ENDOR)[7] and nuclear modulation[8] of spin-echo decay curves enable resolution of superhyperfine splitting associated with ligand nuclear spin coupling to the electron spin.

A final EPR technique to be considered is the Linear Electric Field Effect (LEFE)[9] on the electron spin-echo spectrum; it is a particularly important probe of site symmetry and electronic structure. The high sensitivity of this pulsed EPR technique permits measurement to be made of shifts in g values (proportional to the parameter $\sigma$) caused by an externally applied electric field. This shift is measured as a function of the magnetic field, with the electric field both parallel and perpendicular to the magnetic field. A typical LEFE spectrum is shown at the bottom of Fig. 2; it relates to the continuous-wave EPR signal in the middle of this figure. The signals at low field correspond to the tetragonal complexes in frozen solution with the z axis (hence $g_z$) parallel to the magnetic field. As the magnetic field is increased, molecules with larger projections of $g_\perp$ are probed. From Fig. 2 the average $\sigma$ observed for a tetragonal cupric complex is small and similar for both parallel and perpendicular orientations of the electric field. This is to be expected for a complex which possesses a center of inversion. The electric field has odd parity and must have a zero matrix element for a d orbital ground state. With proteins, larger shift magnitudes thus reflect lower symmetry of the copper site, while orientation dependence of the shift gives insight about odd-parity excited-state mixing into the ground-state wave function (see Sec. III and IV).

## B. Excited States

The excited states are of course studied directly through electronic absorption spectroscopy. The integrated intensity of an absorption band[3], $\int \varepsilon(\upsilon)d\upsilon$, relates theoretically to its oscillator strength (f). This, in turn, is proportional to the square of the electric dipole moment of a transition from the ground state ($\psi_g$) to the excited state ($\psi_e$):

$$f = (4.32 \times 10^{-9})[\int \varepsilon(\upsilon)\, d\upsilon] = (1.085 \times 10^{11})\, \upsilon G\, [\int \psi_e \vec{r}\, \psi_g\, d\tau]^2 \qquad (3)$$

where $\vec{r}$ is the electric dipole operator (odd parity), $\upsilon$ is in $cm^{-1}$, and the integer G depends on the symmetries of $\psi_g$ and $\psi_e$. Although electric dipole selection rules forbid transitions between d orbitals in centrosymmetric molecules, odd parity vibrations of the complex allow these ligand field (d-d) transitions to gain some intensity (vibronic coupling). Three weak transitions to the $d_{x^2-y^2}$ orbital of tetragonal Cu(II) would thus be expected (right side of Fig. 1); these usually appear, depending on field strength of the coordinated ligands, in the red spectral region (~ 600 to 700 nm) as an unresolved broad band with $\varepsilon \approx$ 10 to 50 $M^{-1}$ $cm^{-1}$ (Fig. 3, top).

Figure 4 presents the energy ranges of the composite ligand field bands for tetragonal copper complexes with a representative set of biologically relevant ligands. The energies of these absorption maxima depend somewhat on the tetragonal distortion (T, as defined by Hathaway[11a], which relates equatorial and axial bond lengths). In general, however, as the field strength of the ligands increases, the energies of these transitions will

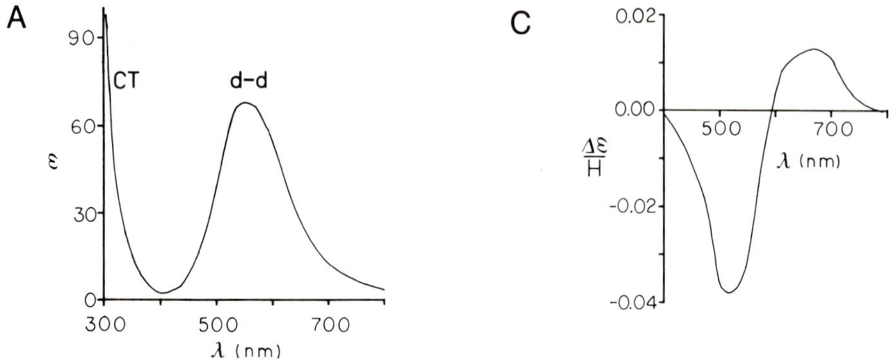


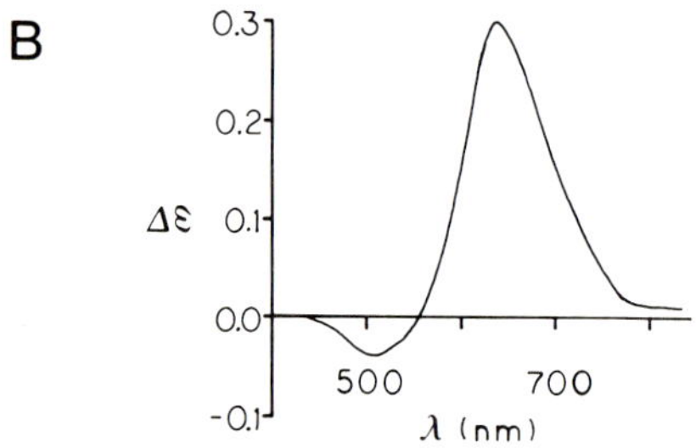


**Fig. 3. A** Optical spectrum of tetragonal tetramine copper(II) complex (adapted from Ref. 11 n). **B** Ligand field circular dichroism spectrum of optically active tetragonal copper(II) complex (adapted from Ref. 11 o). **C** Magnetic circular dichroism of the ligand field region of copper(II) complex in (**A**); scale normalized to 10 K Gauss (adapted from Ref. 11 n)

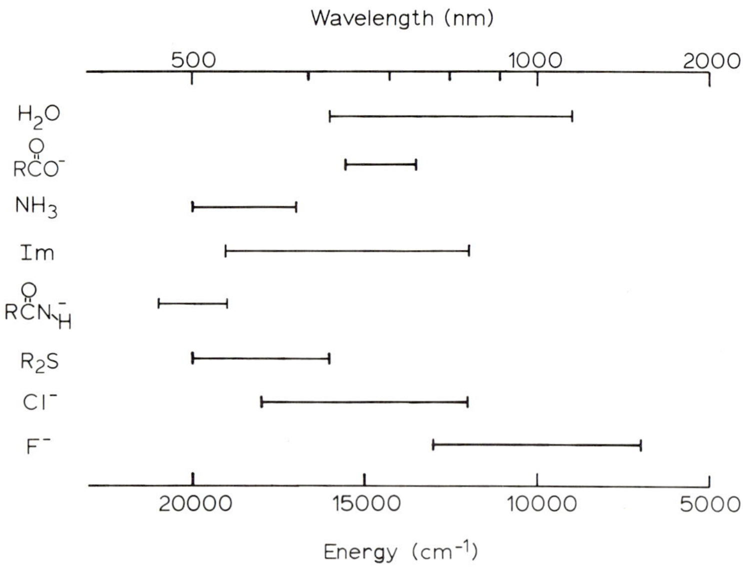


**Fig. 4.** Energy range of ligand field transitions in tetragonal copper(II) complexes with the ligands indicated; ligands (and the corresponding references) are aquo[11b)], carboxylate[5b)], amine[11c)], imidazole[11d, e)], amide[11f)], thioether[11g, h–j)], chloride[11k)], fluoride[11m, l)]

increase. A complex with mixed ligands is expected to produce a spectrum with transitions in an averaged range of energies.

Ligand-to-metal charge transfer (CT) transitions, in which an electron is excited from a predominantly ligand-centered orbital to a predominantly metal-centered orbital (the optical analog of a redox process), are found at higher energy (Fig. 3, top). These are electronic dipole allowed and exhibit much greater intensity ($\varepsilon \approx 10^3$–$10^4\ M^{-1}\ cm^{-1}$); this intensity is proportional to the square of the overlap integral of the orbitals involved. Figure 5 indicates the energy ranges over which charge transfer transitions to copper from biologically relevant ligands in structurally defined tetragonal systems have been observed. In many cases, several low-lying ligand orbitals contribute to the CT spectrum in the visible-UV region. Consequently, many of the bands are due to several overlapping transitions.

Having summarized the types of transitions which are likely to contribute to the absorption spectrum of tetragonal copper (II), it is important to consider methods which probe these transitions and maximize the amount of information which can be ascertained. Linear dichroism[13)] provides the specific orientation of the electric dipole transition moment (Eq. 3) with respect to the copper site. This method is quite valuable in assigning an absorption spectrum (as discussed for Blue Copper proteins in Sect. III) but requires a structurally defined site oriented in a single crystal. A number of other physical

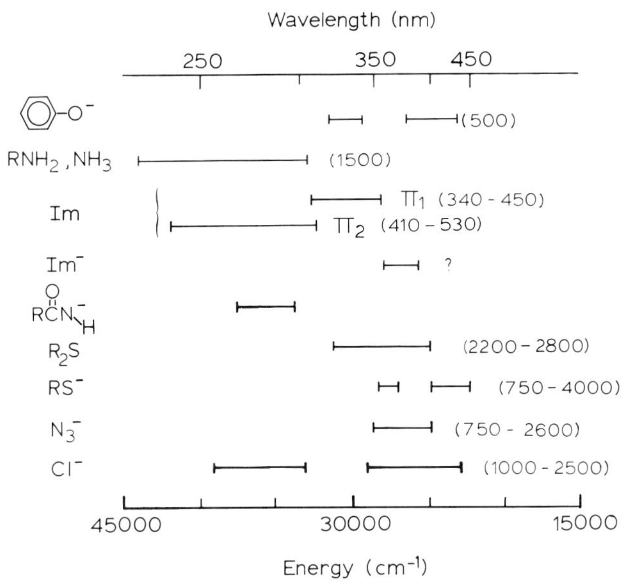


**Fig. 5.** Energy ranges over which CT transitions from the ligands (listed on the *left*) to copper(II) have been observed in model complexes. Values for the extinction coefficients ($\varepsilon$) in $M^{-1}\ cm^{-1}$ are listed to the *right in parentheses*. Ligands (and the corresponding references) are phenolate[12a)], amine[11i, 12b, c)], imidazole[12d, e)], imidazolate[12e)], amide[11f)], thioether[11g, i, 12a)], thiolate[12f, g)], azide[12h, 52)], chloride[45)]. Spectra of both molecules with thiolate ligands had an intense band peaking at approximately 28 000 $cm^{-1}$ with weaker transitions at lower energies (22 000–25 000 $cm^{-1}$). Most $R_2S \rightarrow Cu$ charge transfer transitions were observed in the range of 26 000–31 000 $cm^{-1}$. However, two molecules, each with four thioether sulfur ligands, had transitions at lower energy; this may result from some interaction between the sulfur orbitals

methods, however, provide detailed spectroscopic information even for non-oriented copper sites in solution.

Just as electric dipole selection rules ($\vec{r}$) govern the absorption spectrum, magnetic dipole selection rules ($R_{\vec{r}}$) govern the circular dichroism (CD)[14] spectrum, in which $\Delta\varepsilon = \varepsilon_{left} - \varepsilon_{right}$ is plotted as a function of wavelength. The intensity of a CD band[14a, 3a] is proportional to the rotational strength, R; this is related to the projection of the electric and magnetic dipole moments of a transition:

$$R = (22.9 \times 10^{-40})\left(\int \frac{\Delta\varepsilon}{v}\, dv\right)$$

$$= (4.7 \times 10^{-24}) Im\left(\int \psi_g \vec{r} \psi_e \, d\tau \cdot \int \psi_g R_r \psi_e \, d\tau\right) \quad (4)$$

A number of ligand field and charge transfer transitions are magnetic dipole allowed ($\int \psi_g R_r \psi_e \, d\tau \neq 0$). For example, the transition from $d_{xy}$ to $d_{x^2-y^2}$ in a tetragonal ligand field is strongly magnetic dipole allowed. Thus, in an optically active tetragonal copper complex, this transition dominates the CD spectrum in the ligand field region, as shown in the middle of Fig. 3. The Kuhn anisotropy factor[15], $\gamma = |\Delta\varepsilon/\varepsilon|$, is a measure of the magnetic dipole character and is greater than $\sim .01$ for magnetic dipole allowed transitions.

In magnetic circular dichroism (MCD)[16], an external axial magnetic field is applied to the sample. Even in the absence of natural optical activity, this field induces circular dichroism by splitting ground and excited states in a manner analogous to the EPR experiment. The signals obtained relate quantitatively to the g values of the ground and excited states of a given transition. Qualitatively, the MCD effect provides an additional complementary technique for resolving broad overlapping bands into specific transitions (see Fig. 3, bottom). Moreover, the temperature dependence of the MCD signal is related to the paramagnetism of the ground state and is thus proportional to 1/T (Curie behavior) until low temperature and high magnetic field lead to saturation. For a Kramers doublet this saturation behavior is given[16b] by:

$$\Delta\varepsilon = (\Delta\varepsilon)^{\circ} \tanh(g\beta H/2kT) \quad (5)$$

Therefore, the temperature dependence of an MCD spectrum should clearly indicate which absorption features are associated with paramagnetic centers. The saturation behavior provides an estimate of the ground State g value (Eq. 5) associated with a specific absorption band and can be correlated to a specific EPR signal. This approach should be valuable for probing multicopper oxidase sites (see Sect. V).

The final physical method to be considered here, which allows further probing of an absorption band, is resonance Raman spectroscopy. The excitation laser wavelength is tuned into an absorption band and the vibrations enhanced in the Raman spectrum are detected. Only those vibrational modes associated with distortion of the excited electronic state relative to the ground-state geometry will be resonance enhanced. This method, therefore, not only allows observation of vibrations directly associated with the active site but also provides valuable information on the nature of the excited state. Usually, charge transfer transitions are probed due to the high intensity ($\varepsilon > 500\ M^{-1}\ cm^{-1}$) required for resonance enhancement. These points are well illustrated by reso-

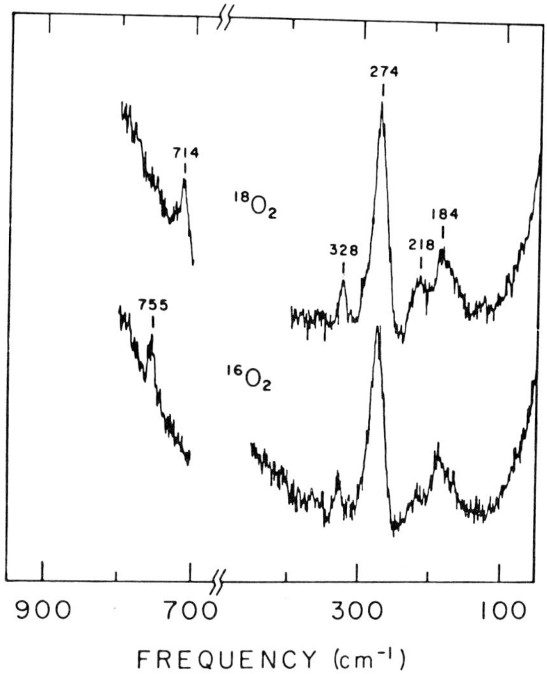


**Fig. 6.** Resonance Raman spectra (200 mW excitation at 363.8 nm) of $^{16}O_2$ and $^{18}O_2$ oxytyrosinase (room temperature; protein concentration = 0.4 mM) (from Ref. 17)

nance Raman studies of the coupled binuclear copper protein site of oxytyrosinase[17)] and oxyhemocyanin[18, 19)]. Their absorption spectra, which will be discussed in Sect. IV, contain two intense bands at $\sim 600$ nm ($\varepsilon \approx 1000$ $M^{-1}$ $cm^{-1}$) and $\sim 350$ nm ($\varepsilon \approx 20{,}000$ $M^{-1}$ $cm^{-1}$). Laser excitation into these features produces an enhanced Raman peak in the intraligand region at $\sim 750$ $cm^{-1}$ and a number of vibrations in the metal-ligand region at about 280 $cm^{-1}$ (Fig. 6). When $^{18}O_2$ is substituted for $^{16}O_2$, the $\sim 750$ $cm^{-1}$ peak shifts to $\sim 710$ $cm^{-1}$, clearly demonstrating that this vibration is an O–O stretch and indicating a peroxide oxidation state ($v_{O_2} = 1580$ $cm^{-1}$; $v_{O_2^-} \approx 1100$ $cm^{-1}$; $v_{O_2^{-2}} \approx 800$ $cm^{-1}$). Further, the fact that the O-O stretch is observed in resonance with these absorption bands indicates that they are due to $O_2^{-2} \rightarrow Cu(II)$ CT transitions.

## *C. Models*

Finally, a comment on the relevance of model studies is warranted. In contrast to heme iron, iron-sulfur proteins, etc., which contain "extrinsic" active sites capable of existing as well-defined entities independent of the protein, all known active sites in copper proteins are "intrinsic"; that is, they are formed only through the intimate interaction of the copper ions with the ligating protein residues. This generates a copper site which is quite different, both in geometry and ligation, from small-molecule copper complexes. It is unrealistic to expect to synthesize a copper complex which is a detailed model for an intrinsic protein site. It is possible, however, to generate small, x-ray crystallographically defined, electronic structural analogues which duplicate specific protein features and which can be studied in detail. In the following sections, spectroscopic studies of appropriately characterized analogues are correlated with the spectral features of specific copper protein sites. Our understanding of the electronic and geometric structure of protein active sites has clearly benefitted from such model studies. As our knowledge of active sites in copper proteins increases, electronic structure analogues will become of even greater importance.

## II. Normal Copper Proteins

Several copper-containing proteins exhibit spectral features similar to those of inorganic copper complexes. The best characterized of these normal copper proteins are Cu-Zn superoxide dismutase (SOD)[20)], galactose oxidase[21)], dopamine-$\beta$-hydroxylase[22)], and the diamine oxidases[23)]. These enzymes differ significantly from one another with respect to quarternary structure, prosthetic groups and reactivity, and thus should each be considered as separate classes of copper sites.

Only Cu-Zn SOD has been defined by x-ray crystallography[24)]. The Cu(II) coordination (Fig. 7, top) is tetragonal and consists of three imidazoles and an imidazolate, which bridges to a tetrahedral zinc(II). While the general EPR and optical features of this active site are similar to those described for normal copper complexes, the protein does modify these spectral characteristics. In particular, the copper site, derived from the interaction of the metal ion and the protein residues, is intrinsically of lower symmetry. The frozen-solution EPR exhibits normal g and A values; however, the low symmetry distinctly manifests itself by a rhombic splitting ($g_x \neq g_y$, $A_x \neq A_y$) of the EPR spectrum (Fig. 7, middle). This appears as a broadening in the $g_\perp$ region of the X band[25)] spectrum and is clearly resolved in the Q band[26)] spectrum ($g_z = 2.26$, $g_y = 2.09$, $g_x = 2.03$, $A_z = 142 \times 10^{-4}\ cm^{-1}$, $A_y = 35 \times 10^{-4}\ cm^{-1}$, $A_x = 52 \times 10^{-4}\ cm^{-1}$; note that all values are for $^{63}Cu$)[27)]. More recent high-resolution x-ray diffraction data[28)] indicate that the copper site has a limited tetrahedral distortion. Single-crystal EPR studies[27)] have determined the orientation of the g and A tensors with respect to the copper site and indicate that $g_x$ and $g_y$ are approximately along the Cu-N(His) bonds. Given this defined orientation of the tensor, LEFE spin-echo experiments would be an ideal probe of this low symmetry site distortion.

Ligand field transitions (Fig. 7, lower) are observed at ~680 nm[25)], consistent with a tetragonal nitrogen ligand field. The large molar extinction coefficient of ~300 $M^{-1}\ cm^{-1}$, however, indicates that the site is of lower symmetry; removal of inversion symmetry enhances intensity of the ligand field transitions. The large increase in d-d band intensity observed upon cooling to 77 K[25)] is consistent with a mechanism in which intensity is induced by a static site distortion (vibronically allowed bands decrease upon cooling). CD and MCD studies[25, 29)] of Cu-Zn SOD (Fig. 7, lower) are limited (i.e., only room temperature MCD has been done), but qualitatively parallel the results for tetragonal complexes in the ligand field region.

Finally, at higher energy are observed ligand to metal charge transfer transitions. Figure 5 indicates that the relatively intense band at 330 nm ($\varepsilon \leqslant 1000$)[30)] should be assigned as a histidine $\rightarrow$ Cu(II) $d_{x^2-y^2}$ CT transition, while the shoulder at 430 nm ($\varepsilon \leqslant 400$)[25)] is probably associated with the bridging imidazolate. This assignment should be confirmed with polarized single-crystal optical studies. Addition of azide to Cu-Zn

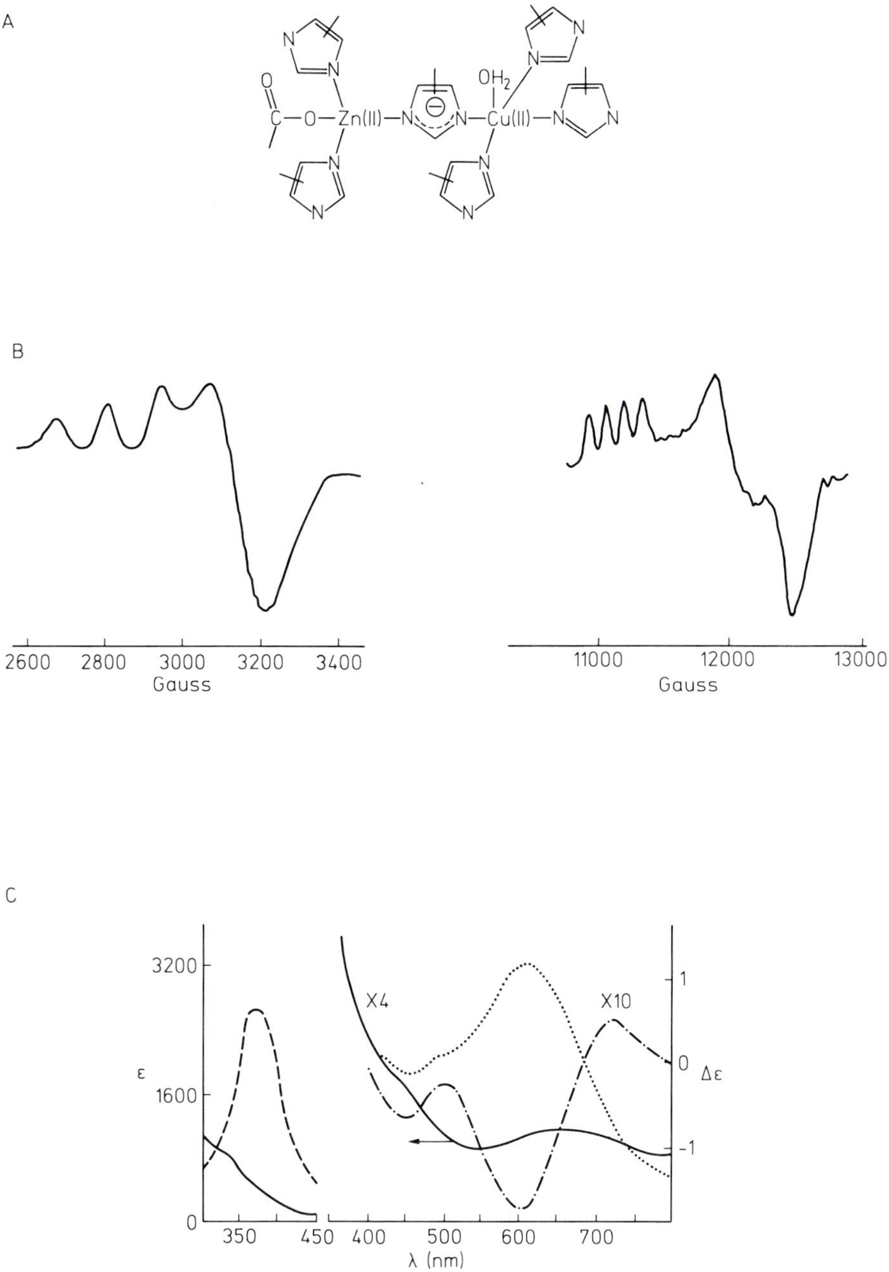


**Fig. 7. A** Structural representation of Cu-Zn superoxide dismutase (SOD) active site (from Ref. 31 c). **B** X-band (*left*) (77 K; $v = 9.2$ GHz) (from Ref. 25) and Q-band (*right*) (173 K; $v \approx 35$ GHz) (from Ref. 26) EPR spectra of bovine Cu-Zn SOD. **C** Optical absorption spectra of native Cu-Zn SOD (—) (Ref. 30), and native + 200-fold excess $N_3^-$ (---) (Ref. 31 a); CD (···) and MCD (–·–·–) (Ref. 29) spectra; intensity of visible absorption spectrum has been expanded by a factor of four; intensity of MCD spectrum has been expanded by a factor of ten and normalized to 10 K Gauss; all samples at room temperature

SOD produces an intense $N_3^- \rightarrow Cu(II)$ CT transition (Fig. 7, lower) at 373 nm ($\varepsilon \approx 2700\ M^{-1}\ cm^{-1}$)[31a]. This energy is consistent with equatorial binding of azide to a tetragonal copper site (Fig. 5). This relatively intense charge transfer transition raises the possibility of a UV resonance Raman spectroscopic investigation of competitive inhibitor[31b] (i.e., $N_3^-$) binding to this active site.

# III. Blue Copper Proteins

## A. Introduction

Stellacyanin, the plastocyanins, and the azurins are the most widely studied copper-containing metalloproteins of the next active-site class, the Blue Copper sites. These proteins, which generally appear to be involved in redox chemistry, have quite unique spectral features[32, 33]. The potential for complementary interaction between inorganic spectroscopy and protein crystallography is well demonstrated by the roles that they have played in generating fairly detailed geometric and electronic structural pictures of the Blue Copper metal centers.

Spectral features associated with the Blue Copper site are clearly different from those of normal copper complexes (Fig. 8). While $g_{\parallel} > g_{\perp} > 2$ is found in the EPR spectra of both normal copper complexes and the Blue Copper site, the latter exhibit a much smaller copper parallel hyperfine splitting ($A_{\parallel} < 70 \times 10^{-4}\ cm^{-1}$). The axial nature of these protein EPR signals is quite surprising since a rhombic signal would be expected from a copper geometry distorted by the protein (see discussion in Sect. II). However, the most striking spectral feature of the Blue Copper proteins, and the source of their name, is the very intense absorption band in the red spectral region, peaking at about 600 nm with an $\varepsilon$ on the order of 5000 $M^{-1}\ cm^{-1}$.

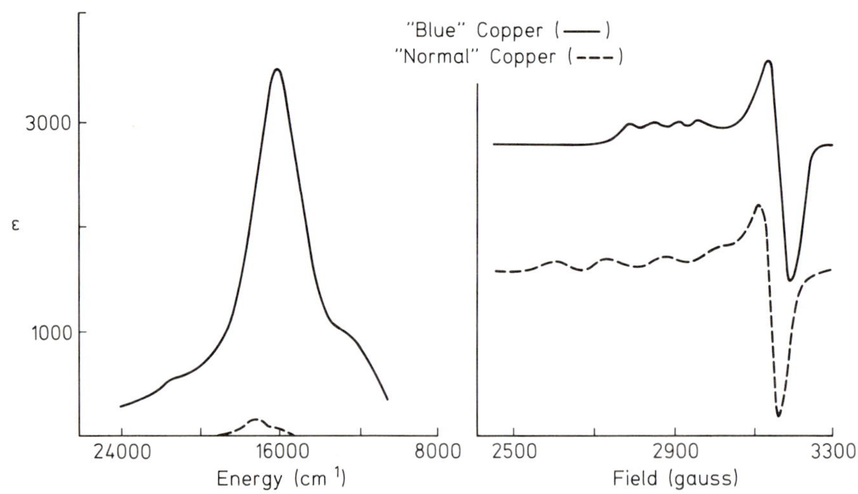


**Fig. 8.** Optical (*left*) and EPR (*right*) spectra of a Blue Copper protein (*solid line*) and a tetragonal copper site (*dashed line*) (from Ref. 1)

## B. Geometric Structure

The initial goal of inorganic spectroscopic studies on the Blue Copper proteins was to understand their spectral features and thereby generate active site structural models. A significant advance in this direction came with the extension of spectral studies into the near-IR[34, 35], using the newly developed infrared circular dichroism instrumentation of Stephens[36] (Fig. 9). The point of these studies was to look for electronic transitions in a region where the absorption spectrum is normally dominated by protein vibrations. Since the magnetic dipole character of the ligand field transitions is much greater than that of vibrations, these d-d transitions would be expected to be more intense, relative to the protein vibrations, in the CD spectrum. Three bands, previously unobserved, were found in the spectral region between 5000 and 12 000 $cm^{-1}$ and assigned as ligand field transitions. These transitions were used to determine the geometry of the active site by means of the ligand field theoretical calculation described in Fig. 10. This figure shows changes of the d orbital *state* energies in a four-coordinate Cu(II) complex as its geometry is varied from square planar ($\beta$ = 90°) to tetrahedral ($\beta$ = 54.74°). (The right-hand side of Fig. 1 would correspond to the right-hand side of Fig. 10 in the limit of infinite axial bond length.) Near the tetrahedral limit, all of the transitions are predicted to be much lower in energy than in a square planar complex. As predicted by ligand field theory, this shift of the d-d bands to lower energy has been verified experimentally for a series of $CuCl_4^{2-}$ complexes with several values of the angle $\beta$ (Ref. 35, pp. 174–5). The energies of the transitions observed in near-IR CD spectra of the Blue Copper proteins were fit to this scheme with a geometry which is close to tetrahedral ($\beta$ = 60–61°). Further, the EPR signals ($g_{\parallel} > g_{\perp} > 2$) required that the half-occupied orbital of the ground state be $d_{x^2-y^2}$. The axial nature of the EPR spectrum indicated then that the tetrahedron is distorted either into a $D_{2d}$ or $C_{3v}$ geometry. In the $D_{2d}$ distortion, the angles of each of the bonds

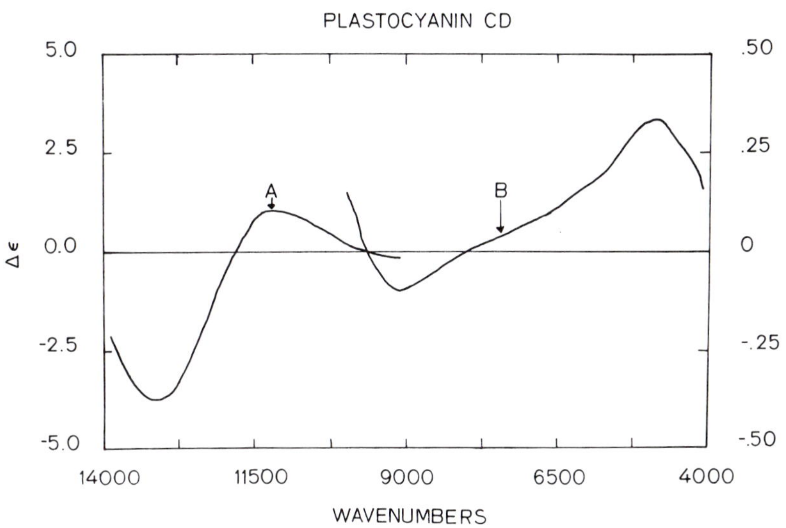


**Fig. 9.** Near-infrared circular dichroism spectrum of plastocyanin in $D_2O$ at 290 K. Spectrum A corresponds to scale on *left*. Spectrum B corresponds to scale on *right* (from Ref. 35)

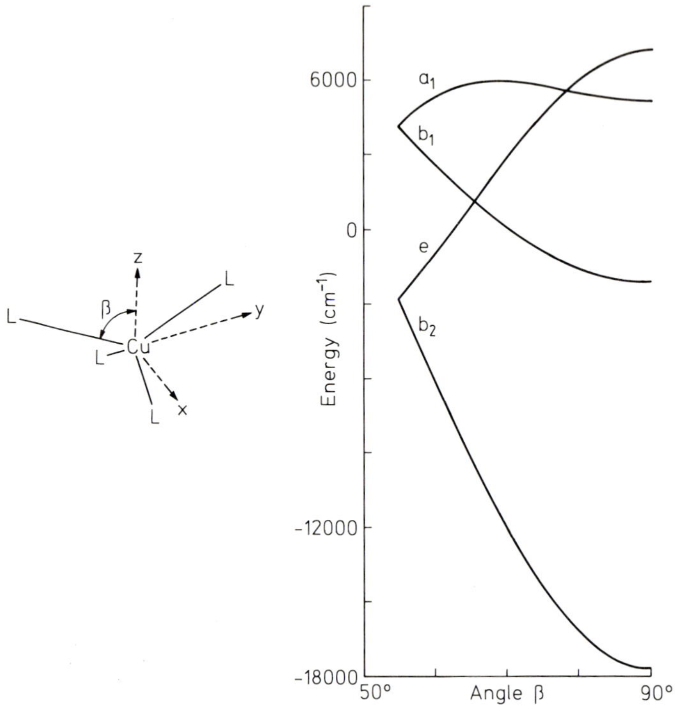


**Fig. 10.** Theoretical shifts in the energies of the d orbital *states* of a $d^9$ ion with four equivalent ligands (Ds = 765 $cm^{-1}$; Dt = 444 $cm^{-1}$) as the structure of the site is varied from tetrahedral ($\beta$ = 54.7°) to square planar ($\beta$ = 90°). The symmetry designation (and the corresponding half-filled d orbital) of each state are: $a_1(d_{z^2})$; $b_1(d_{x^2-y^2})$; $e(d_{xz}, d_{yz})$; $b_2(d_{xy})$ (from Ref. 35)

with the z axis all increase or all decrease (one limit of this distortion is a square planar complex). To produce a site with $C_{3v}$ symmetry, one of the bonds must be lengthened or shortened. It is generally believed[37, 38] that mixing of $p_z$ character into the $d_{x^2-y^2}$ ground state is necessary to decrease A values in tetrahedral copper complexes. Several copper(II) complexes with approximately $D_{2d}$ distorted tetrahedral symmetry, which permits this mixing, exhibit very small hyperfine parameters[37a–c] while a copper(II) complex with a compressed trigonal site, which does not allow $p_z$ mixing, shows a large copper hyperfine splitting [37e].

Based on these studies, the low value of the parallel hyperfine splitting in conjunction with the axial EPR spectra appeared to support the $D_{2d}$ effective axial symmetry. Additionally, since all the ligand field transitions were observed below 800 nm, the dominant spectral features in the visible region had to be assigned to charge transfer transitions. Titrations of the Blue Copper proteins with $HgCl_2$ and parachloromercuribenzoate[39], as well as ESCA studies[40] and electronic spectral studies of cobalt(II) substituted Blue Copper proteins[41], indicated that cysteine was probably a ligand; this suggested that the intense low-energy features at ~600 nm were attributable to cysteine-to-copper charge transfer transitions.

The crystal structure of plastocyanin, determined to a resolution of 1.6 Å by Freeman[42] (Fig. 11), confirmed the near-tetrahedral geometry of this Blue Copper site and

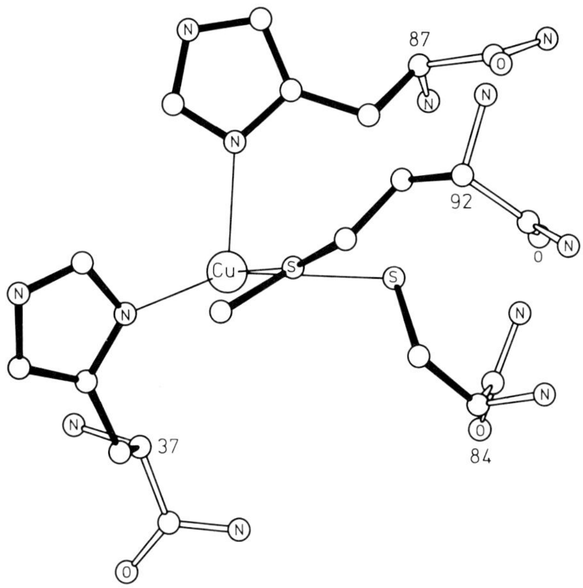


**Fig. 11.** The Blue Copper site in plastocyanin as determined by x-ray crystallography. Ligands (and copper-ligand bond lengths) are histidine 37 (2.04 Å), cysteine 84 (2.13 Å), histidine 87 (2.10 Å), and methionine 92 (2.90 Å) (from Ref. 42)

the presence of a cysteine ligand which provided a short (2.1 Å) thiolate-copper bond. It also showed that the copper had two imidazole ligands (consistant with earlier paramagnetic NMR studies[43]) and demonstrated the existence of a long (2.9 Å) copper-methionine bond. However, a number of questions about the electronic structure of the copper site remained to be answered. These questions concerned (1) the detailed nature of the charge transfer spectrum (at least five bands are present, *vide infra*); (2) contributions of the different ligands to bonding interaction with the copper (in particular, the quite long Cu–Met and short Cu–Cys bonds); (3) the relationship between the unique axis (defining the $d_{x^2-y^2}$ orbital which is approximately normal to $g_z$ and involved in electron transfer) and the structure of the site, and (4) the origin of this "axial symmetry" in a distorted copper site of clearly low symmetry.

## C. Excited States

In order to understand the charge transfer features of the Blue Copper site, the variable-temperature optical absorption, room-temperature circular dichroism, and magnetic circular dichroism spectra of plastocyanin, stellacyanin, and azurin were studied[35]. As can be seen for plastocyanin in Fig. 12, the relative intensities (and signs, in the case of CD and MCD) of these transitions vary among the different types of spectra. This is a result of the difference in selection rules for absorption, CD, and MCD spectra, as mentioned in the Introduction. A careful comparison of the three types of spectra and the absorption bandshape temperature dependence (see moment analysis in Ref. 35, pp. 176–177)

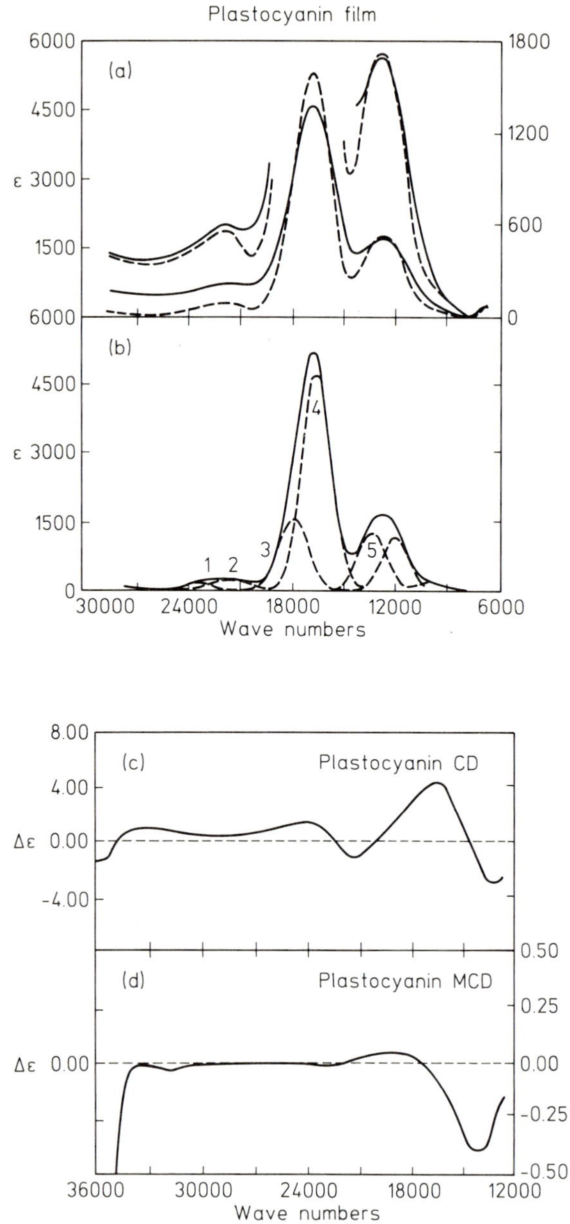


**Fig. 12 a–d.** Visible and near-IR spectra of plastocyanin (adapted from Ref. 35): **a** Absorption spectrum of films at 270 K (*solid line*) and 35 K (*dashed line*); *lower curves* refer to left-hand scale, *upper curves* to right-hand scale. **b** Gaussian resolution of the 35 K absorption spectrum. **c** Circular dichroism spectrum of plastocyanin in pD = 6 deuterated phosphate buffer at 290 K. **d** Magnetic circular dichroism spectrum of plastocyanin in deuterated phosphate buffer at 290 K

made possible Gaussian analysis of the optical spectrum of each protein. The results are indicated by dashed lines in part B of the figure. Note that a band at 18 000 $cm^{-1}$, which is almost completely obscured in the optical spectrum by the more intense band at 16 500 $cm^{-1}$, is clearly visible as a shoulder on the 16 500 $cm^{-1}$ band in the CD spectrum. Thus, at least five CT transitions (bands 1–5) are observed in the optical spectrum at energies higher than the highest energy d-d band (from the ligand field calculation) at 11 200 $cm^{-1}$.

In our initial assignment[34, 35], two transitions were attributed to charge transfer from the cysteine sulfur to the copper. This assignment was based on a model in which only the S–Cu bond was considered, leading to a hybridization of the sulfur valence orbitals along this bond (Fig. 13, left). In this model, the $S(\alpha s + \beta p_z)$ orbital would $\sigma$-bond to the Cu and be stabilized to deeper binding energy relative to the degenerate $p_x$ and $p_y$ orbitals, which would be involved only in $\pi$-bonding. The higher-energy $\sigma$ $S(\alpha s + \beta p_z) \rightarrow Cu(d_{x^2-y^2})$ transition would be only electric dipole allowed. It would be intense in absorption because of good overlap between the $\sigma$ orbital on the sulfur and the $d_{x^2-y^2}$ orbital on the copper but would have a small Kuhn anisotropy factor. The $\pi$ $S(p_x, p_y) \rightarrow Cu(d_{x^2-y^2})$ transition would be magnetic dipole allowed but relatively weak in absorption (overlap between the orbitals involved is small). Therefore, this transition would have a large value of $\gamma$. The assignments of the transitions in plastocyanin at 13 000 $cm^{-1}$ (band 5) to $\pi S \rightarrow d_{x^2-y^2}$ ($\gamma \sim 0.01$)[35] and at 16 500 $cm^{-1}$ (band 4) to $\sigma S \rightarrow d_{x^2-y^2}$ ($\gamma \sim 0.001$)[35] were supported by this model. In addition, the transition energy and Kuhn anisotropy factor of band 2 ($\sim$22 000 $cm^{-1}$, $\gamma \sim 0.01$)[35] indicated it was due to a $\pi$-type charge transfer transition, probably originating from a $\pi$ orbital on a histidine ring ($\pi$N).

Further insight into the charge transfer features was recently obtained[44] through single-crystal polarized optical studies of plastocyanin. In these experiments, absorption spectra were obtained (Fig. 14) with polarized light which had its electric dipole direction oriented parallel and perpendicular to the crystal $\vec{a}$ axis (Fig. 15). These spectra were

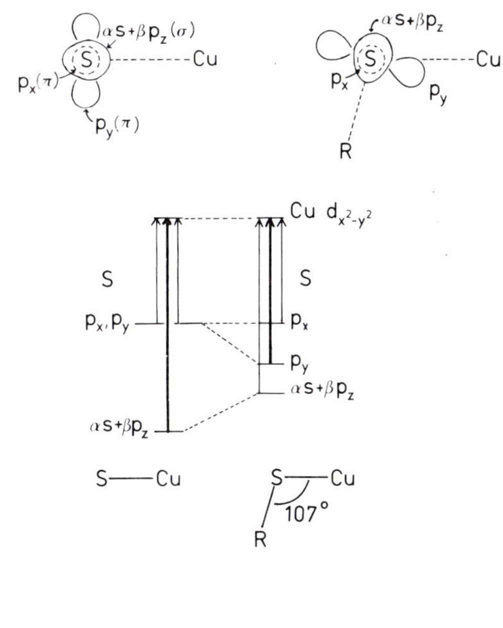


**Fig. 13.** Two Cys sulfur valence orbital hybridization schemes and the corresponding energy-level diagrams. *Left:* Considering only the sulfur and copper atoms. *Right:* Hybridization along the Cys Cu–S axis. The R–S–Cu angle, as defined by x-ray crystallography, is 107°. Vertical arrows in the energy-level diagrams designate charge transfer transitions. In diagrams of bonding geometry, the z direction is defined as along the R–S bond, y is perpendicular to z and in the plane of the page, and x is perpendicular to the plane of the page

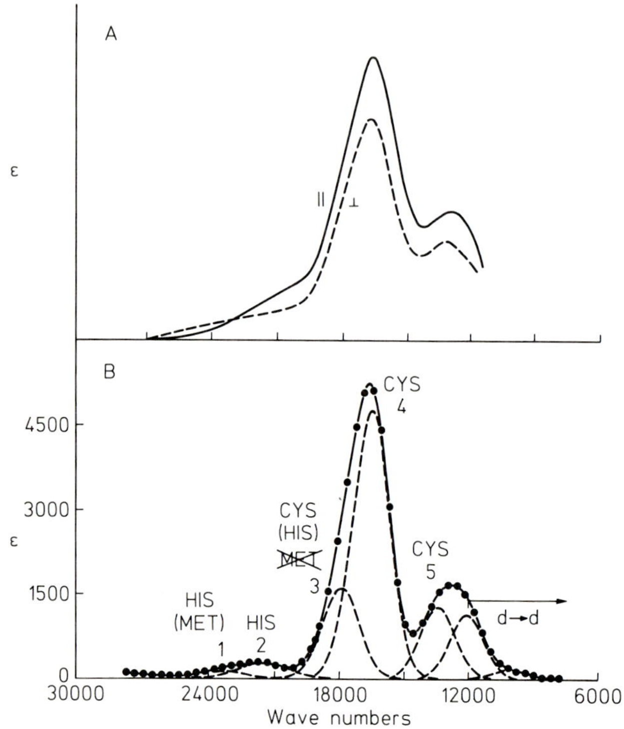


**Fig. 14. A** Room-temperature optical spectrum of a single crystal of plastocyanin obtained with light incident on the (0, 1, 1) face and polarized parallel (*solid line*) and perpendicular (*dashed line*) to ã (from Ref. 44). **B** Gaussian resolution of the 35 K visible absorption spectrum of a plastocyanin film with suggested assignments; the symbols (•) represent the experimental absorption spectrum

interpreted using the approximation (based on overlap arguments) that the transitions are polarized along the metal-ligand axes. Taking the symmetry of the unit cell (which contains four molecules) into consideration and appropriately projecting the electric dipole vector onto all metal-ligand axes, the polarization ratios of charge transfer transitions from each of the four ligands and from the two hisitidines considered together were calculated. These are listed in Table 2 along with the experimentally observed polarization ratios for bands 1 to 5. Also useful in the assignment of these spectra were the results of studies on tetragonal spectral analogues (Fig. 5). Correlations between CT transitions in tetragonal ($D_{4h}$) and distorted tetrahedral ($D_{2d}$) sites have been demonstrated in a recent detailed single-crystal study of $CuCl_4^{2-}$ complexes[45]. Charge transfer transitions in the $D_{2d}$ systems were shown to be as much as 5000 $cm^{-1}$ lower in energy than the corresponding transitions in $D_{4h}$ sites. Although these shifts might be expected to be predominantly due to changes in the energy of the copper $d_{x^2-y^2}$ orbital, it was found that ligand-ligand repulsions and overlap differences are important in determining the shifts of individual transitions. In studies of copper-imidazole and copper-pyrazole complexes, it has been shown that CT transitions can shift to lower energy by up to 10000 $cm^{-1}$ as the symmetry changes from tetragonal to nearly tetrahedral[12d, e]. In addition, these

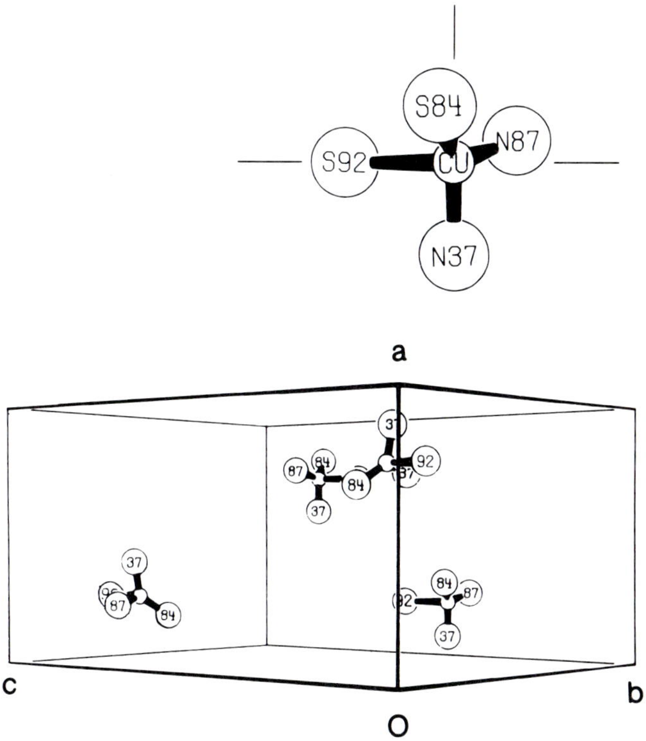


**Fig. 15.** Plastocyanin unit cell projected on the (0, 1, 1) plane, showing the positions of the four symmetry-related Cu atoms at their first coordination shells. The Cu-ligand bonds are drawn on twice the scale of the unit cell (*lower*). Expanded view of the Cu atom and ligands (*upper*) (from Ref. 44)

**Table 2.** Polarization Rates

| Band no. | $v_1$ cm$^{-1}$ | Observed $I_{\|}/I_{\perp}$[a] |
|---|---|---|
| 1 | 23340 | 0.0 –0.4 |
| 2 | 21390 | 1.6 –2.3 |
| 3 | 17870 | 1.1 –1.8 |
| 4 | 16490 | 1.27–1.32 |
| 5 | 13350 | 1.2 –1.6 |
| **Ligand** | | **Caculated $I_{\|}/I_{\perp}$** |
| Met(92) | | 0.0 |
| His(37) | | 84.23 |
| His(87) | | 0.12 |
| His(37) + His(87) | | 2.49 |
| Cys(84) | | 0.96 |

[a] Ranges of values were determined from Gaussian analyses of the two sets of spectra

studies have indicated that two $\pi$ orbitals can be involved in charge transfer transitions from a given histidine. Transitions from the $\pi_2$ orbitals should be 4000 $cm^{-1}$ or more higher in energy than transitions from the $\pi_1$ levels; it appears that both transitions should have the same polarizations.

The mixed polarizations of bands 4 and 5 in the single-crystal spectrum support their assignment as Cys $\rightarrow$ Cu charge transfer. Similarly, the assignment of band 2 to a His $\rightarrow$ Cu charge transfer is reinforced by its strong parallel polarization; the observed polarization ratio most closely fits the theoretical ratio for transitions from both histidines at the same energy. The polarization ratio of band 1 (stronger in perpendicular) shows that this transition originates from either the methionine or histidine (87). Studies of spectral analogues[11g, i)] suggest that band 1 is probably too high in energy to be due to methionine. However, assignment of band 1 to a transition from His (87) requires band 2 to be due to a transition from His (37) alone. This assignment, though plausible, results in somewhat poorer agreement between the observed and predicted polarization ratios for band 2.

Polarization ratios clearly indicate that earlier assignments[35, 46)] of band 3 as a Met $\rightarrow$ Cu charge transfer are unlikely, as it is *not* perpendicularly polarized. This conclusion is also supported by the high intensity of band 3, which is inconsistent with the length of the Cu–Met bond and poor overlap of the Cu $d_{x^2-y^2}$ orbital with the S(Met) orbitals (*vide infra*), and by the presence of this band in the CD spectrum of stellacyanin[35)], which contains no methionine[47)]. The observed polarization ratios are consistent with the assignment of this band to a charge transfer transition involving either the cysteine or possibly both histidines. However, spectral analogue studies indicate[12d, e)] that the energy of band 3 is too low for a His $\rightarrow$ Cu charge transfer, even for tetrahedral symmetry. Therefore, band 3 has been suggested[44)] to be a third Cys $\rightarrow$ Cu charge transfer transition.

This led to a consideration of the orientation of the Cys residue in a detailed description of its bonding to the Cu site. As mentioned above, only two CT bands initially had been proposed to involve cysteine. This was based on a model in which the axis of hybridization of the sulfur orbitals was the S–Cu bond. However, as the C–S bonding interaction in the Cys residue is stronger than the S–Cu bond, it is more appropriate to view the S valence orbitals as hybridized along the C–S bond (Fig. 13). The crystal structure has shown[42)] the Cys C–S–Cu angle to be 107°. Thus, the degeneracy of the sulfur $p_x$ and $p_y$ orbitals should be split by bonding to copper, with the energy of the $p_y$ orbital lowered by its greater participation in the bonding (see Fig. 13, right). Three transitions can therefore originate from the cysteine, with the central one having the greatest intensity since it involves the $\sigma$ bonding sulfur $p_y$ orbital and therefore has the most overlap with the copper $d_{x^2-y^2}$ orbital. Although detailed single-crystal spectral studies on a copper-thiolate complex are required to obtain a good estimate for the energy of a cysteine $S(\alpha s + \beta p_z) \rightarrow Cu(d_{x^2-y^2})$ charge transfer transition, three cysteine-to-Cu charge transfer transitions are observed in a 5000 $cm^{-1}$ region for a spectral analogue[12g)]. Further characterization of these transitions through detailed single-crystal optical absorption experiments on spectral analogues and self consistent field X$\alpha$ scattered wave (SCF-X$\alpha$-SW) bonding calculations on the Blue Copper protein site are presently underway.

## D. Ground State

In addition to polarized optical studies, single-crystal EPR experiments have been performed on plastocyanin[44]. In an axial system, each individual molecule contributes to the EPR spectrum with a g value[4a] given by:

$$g = (g_{\parallel}^2 \cos^2\theta + g_{\perp}^2 \sin^2\theta)^{1/2} \tag{6}$$

where $\theta$ is the angle between the magnetic field and the molecular $\vec{z}$ axis. In a powder or frozen solution there is a distribution of molecules representing all values of $\theta$ from 0° to 180°. In a single crystal at a given orientation, however, only a fixed $\theta$ value for each molecule in the unit cell is represented. In the plastocyanin crystal (space group $P\,2_12_12_1$) all of the molecules are required to be magnetically equivalent (i.e., have equal values of $\theta$) only when the magnetic field is directed along a crystal axis. As shown in Fig. 16, an EPR signal similar to $g_{\parallel}$ was obtained when the magnetic field was parallel to the $\vec{c}$ axis and a $g_{\perp}$ type signal was obtained with the magnetic field along $\vec{b}$. By simulating the EPR spectra for six rotations with four molecules in the unit cell, it was found that $g_z$ is oriented 8° ± 2° off the $\vec{c}$ axis in the bc plane and 5° ± 3° out of the bc plane. This placed $g_z$ in the vicinity of the methionine. However, because four molecules are contained in the unit cell, the EPR experiment alone could not indicate which one of four $g_z$ orientations was associated with a specific molecule in the unit cell. In order to determine the correct orientation of $g_z$, as well as to answer some more general questions about the electronic structure of the Blue Copper site, a ligand field calculation of this site was performed. The d orbital energy levels and wavefunctions were calculated using the method of Companion and Komarynsky[48]. In this strong field method, no simplifying assumptions about site symmetry are made and each element of the secular determinant is calculated using the real d orbitals as the basis set. Each ligand is treated as a point charge; its electrostatic interaction with the metal d electrons is viewed as a perturbation of the d energy levels and depends on the identities of the metal and the ligand and on the metal-ligand bond length. Initially, ligand field parameters were obtained from copper model complexes[11d, e, g, h, i, j, 12g] (see Fig. 4). These were extrapolated to the Blue Copper site, adjusted to fit the Blue Copper spectral features, and extrapolated back to the spectral analogues, where their accuracy was verified (see Table 3). As a result of these calculations, ligand field energy levels (Fig. 17) and the ground state EPR g tensor were obtained. The direction of $g_z$ was determined to be 5° off the methionine-copper bond, placing the half-occupied $d_{x^2-y^2}$ orbital, which is involved in electron transfer, less than 15° above the remaining three ligands (Fig. 18). These calculations also provided some fairly accurate bonding parameters; they show that the imidazoles make a somewhat larger contribution to the ligand field than does the cysteine (short Cu–S bond), and that the methionine, with its long Cu–S bond, makes a very weak *but real* contribution to the ligand field of the site. The nature of the distortion from tetrahedral symmetry was also investigated. As mentioned previously, two distortions from a tetrahedral geometry ($D_{2d}$ and $C_{3v}$) retain the axial symmetry of the site. By removing the large rhombic distortion present in the calculated energy levels, it was found that the spacing and order of these levels is only consistent with a $C_{3v}$ geometry where one bond is elongated and the three other ligands move closer to the newly defined xy plane. The rhombic splitting, which is associated with three ligands in the xy plane, still generates an axial EPR signal since it

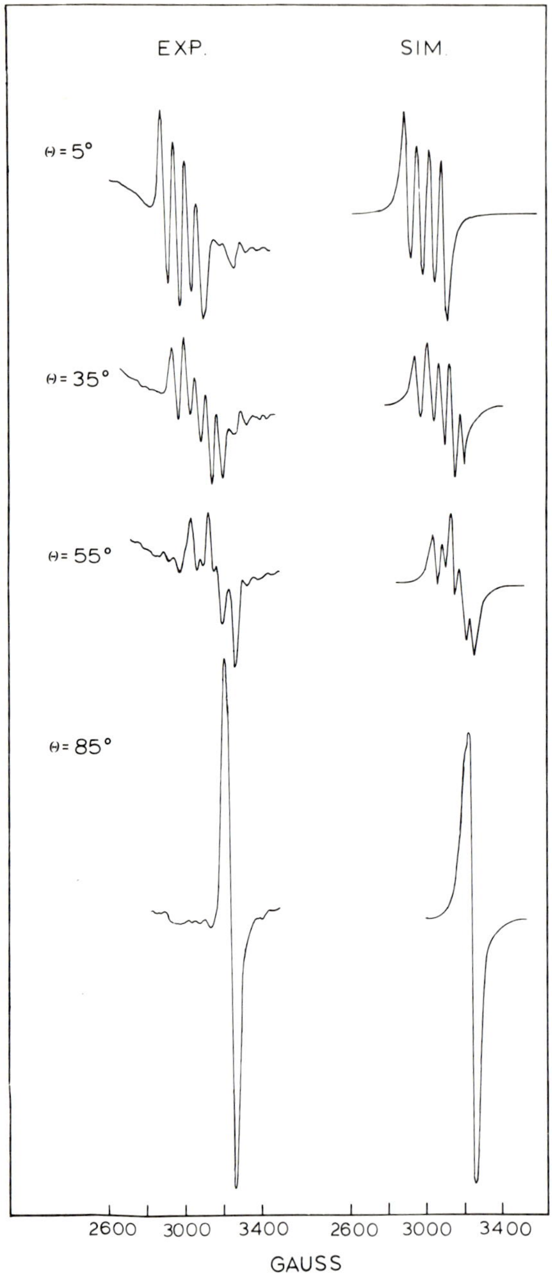


**Fig. 16.** Experimental and simulated EPR spectra of a single crystal of plastocyanin with the crystal $\vec{a}$ axis perpendicular to the plane of the magnetic field; $\theta$ is the angle between the crystal $\vec{c}$ axis and the magnetic-field direction. The simulation was performed with four molecules in a unit cell, $g_z$ 8° off $\vec{c}$ in the $\overline{bc}$ plane and 5° out of the $\overline{bc}$ plane, and $g_z$ and $A_z$ colinear (from Ref. 44)

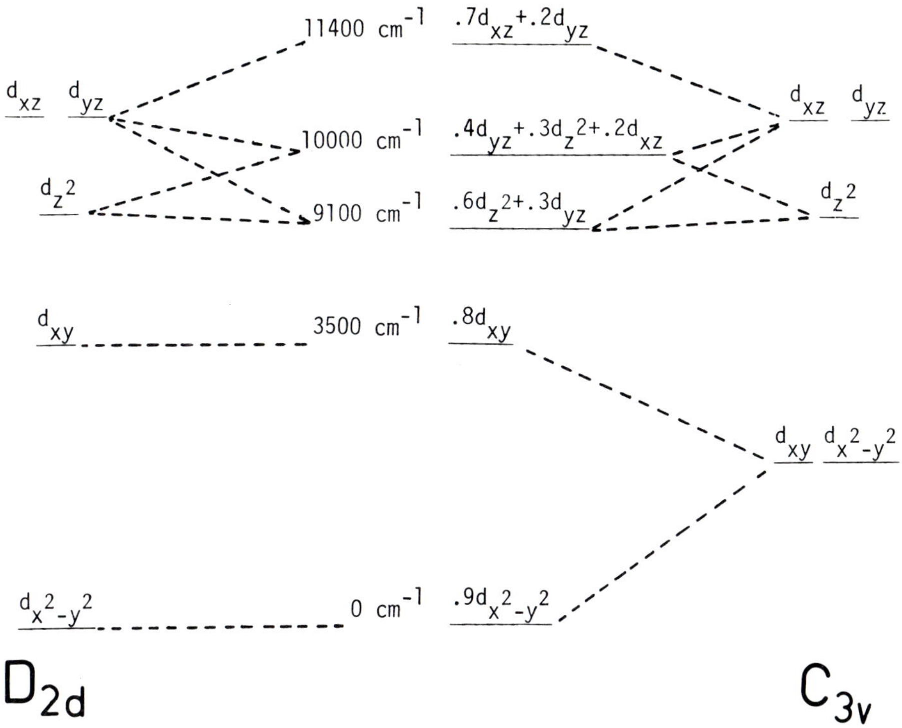


**Fig. 17.** Ligand field energy-level diagram calculated for plastocyanin. Center contains energies and wavefunctions (squared coefficients of leading terms) of the copper site. Energy levels determined after removing the rhombic distortions to give $D_{2d}$ and $C_{3v}$ symmetries are shown in the left and right columns, respectively (from Ref. 44)

**Fig. 18.** Electronic structural representation of the plastocyanin active site. Note that the Cu–S(Met) bond is only 5° off $g_z$, while the bonds from Cu to N(His 37), N(His 87), and S(Cys) are all less than 15° below the xy plane (from Ref. 44)

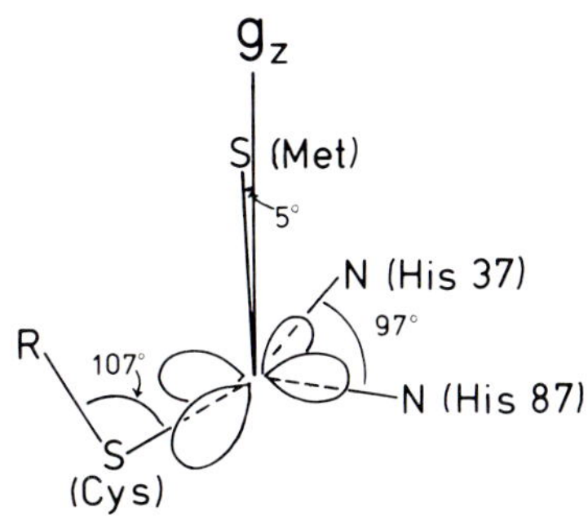


produces a large splitting of $d_{x^2-y^2}$ and $d_{xy}$, but only a very small splitting of $d_{xz}$ and $d_{yz}$. (The magnitude of this latter splitting determines the presence or absence of a rhombic EPR spectrum, in accordance with Eq. 1b.) Further, considering the results described earlier of EPR studies on $D_{2d}$ and $C_{3v}$ copper complexes, it appears that the rhombic distortions of this elongated $C_{3v}$ Blue Copper site may be necessary to produce small A

**Table 3.** Ligand Field Parameters for Plastocyanin and Appropriate Model Complexes

| Model Ligand | Bond length (Å) | $\alpha_2$ (cm$^{-1}$)[a] | $\alpha_4$ (cm$^{-1}$) | Plastocyanin Bond length (Å) | $\alpha_2$ (cm$^{-1}$) | $\alpha_4$ (cm$^{-1}$) |
|---|---|---|---|---|---|---|
| Hys | 2.01[b, c] | 14,000 | 7,500 | 1.97 | 14,915 | 8,334 |
| Cys | 2.36[d] | 12,000 | 6,000 | 2.10 | 10,000 | 7,500 |
| Met | 2.30[e](2.37)[f] | 13,400 | 6,700 | 2.90 | 4,000 | 1,000 |

[a] $\alpha_\ell \equiv Ze^2 \int_0^\infty R_{3d}^2 [r_<^\ell / r_>^{\ell+1}] r^2 dr$ where $\ell = 2,4$ (see Ref. 48)

[b] Ligand field fit to tetrakis(imidazole)dinitratocopper(II) model complex spectrum (Ref. 11 d) included $\alpha_2$(Cu–O) = 3,500 cm$^{-1}$ and $\alpha_4$(Cu–O) = 2,400 cm$^{-1}$ (Cu–O bond length = 2.6 Å) for axial nitrates

[c] Spectral fit for hexakis(imidazole)copper(II) nitrate (Ref. 11 e) included $\alpha_2$ = 13,250 cm$^{-1}$ and $\alpha_4$ = 6,850 for Cu–N bond length of 2.05 Å and $\alpha_2$ = 6,550 cm$^{-1}$ and $\alpha_4$ = 2,110 cm$^{-1}$ for Cu–N bond length of 2.59 Å

[d] Spectral fit of [rac-5,7,7,12,14,14,-hexamethyl-1,4,8,11-tetraazocyclotetradecane]copper(II)*o*-mercaptobenzoate hydrate, [Cu(tet b)(*o*-$SC_6H_4CO_2$)] · $H_2O$ (Ref. 12 g) included $\alpha_2$ = 13,350 cm$^{-1}$ and $\alpha_4$ = 6,520 cm$^{-1}$ for Cu–N = 1.99 Å, $\alpha_2$ = 12,700 cm$^{-1}$ and $\alpha_4$ = 6,020 cm$^{-1}$ for Cu–N = 2.03 Å, $\alpha_2$ = 10,940 cm$^{-1}$ and $\alpha_4$ = 4,670 cm$^{-1}$ for Cu–N = 2.13 Å and $\alpha_2$ = 10,040 cm$^{-1}$ and $\alpha_4$ = 4,050 cm$^{-1}$ for Cu–N = 2.19 Å. These values were determined from a spectral fit of $Na_4Cu(NH_3)_4\{Cu(S_2O_3)_2\}_2$, $H_2O$ and $Na_4Cu(NH_3)_4\{Cu(S_2O_3)_2\}_2$, $NH_3$ (Ref. 11 c)

[e] Spectral fit of 1,4,8,11-tetrathiacyclotetradecane copper(II) diperchlorate (Refs. 11 g and 11 h) included $\alpha_2$ = 3,180 cm$^{-1}$, $\alpha_4$ = 2,040 cm$^{-1}$ for Cu–O = 2.65 Å

[f] Spectral fit of bis[$\beta$-methylmercaptoethylamine]copper(II) diperchlorate (Refs. 11 i and 11 j) included $\alpha_2$ = 13,710 cm$^{-1}$, $\alpha_4$ = 6,820 cm$^{-1}$ for Cu–N = 1.98 Å and $\alpha_2$ = 3,370 cm$^{-1}$, $\alpha_4$ = 2,250 cm$^{-1}$ for Cu–O = 2.60 Å

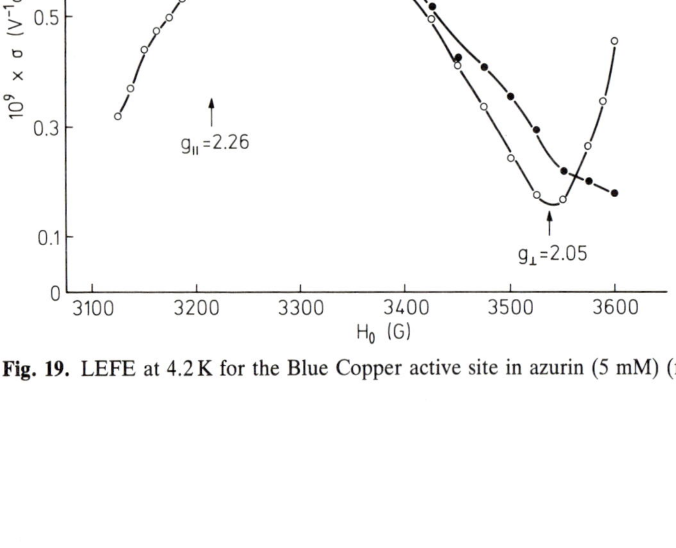


**Fig. 19.** LEFE at 4.2 K for the Blue Copper active site in azurin (5 mM) (from Ref. 49)

values; these distortions would permit the mixing of s as well as p character into the ground state $d_{x^2-y^2}$ wavefunction. However, further insight into the origin of the small parallel hyperfine splitting requires a greater knowledge of the wavefunction of this ground state. This may come about through the SCF-X$\alpha$-SW calculations presently underway for the plastocyanin copper site.

An additional probe of interconfigurational mixing into the ground state wavefunction is the Linear Electric Field Effect (LEFE). In this spectral technique, the Blue Copper site shows[49] behavior quite different from that of centrosymmetric tetragonal copper complexes (Fig. 2). In particular, a large $\sigma$ is observed, consistent with the low symmetry of the Blue Copper site (Fig. 19). Further, the maximum shift is found with $\vec{E} \perp \vec{H}$ at the magnetic field corresponding to $g_{\parallel}$. Since the orientation of the g tensor with respect to the active site has been determined by the single-crystal EPR experiments described earlier, orientation dependence of the large LEFE shifts corresponds with the xy plane containing the Cys–Cu bond. This indicates a large mixing of the Cys(S) $\rightarrow$ Cu($d_{x^2-y^2}$) charge transfer (or $p_x$ and $p_y$) excited state into the ground state wavefunction.

# IV. Coupled Binuclear Copper Proteins

## A. *Introduction*

The next class of copper proteins also exhibits spectral features quite different from those of normal tetragonal copper(II). These contain a coupled binuclear copper site and include hemocyanin, the oxygen-binding protein of molluscs and arthropods, and tyrosinase, an enzyme capable of hydroxylating monophenols (mono-oxygenase activity) and oxidizing ortho-diphenols (oxidase activity)[2, 50]. Mollusc hemocyanin and tyrosinase also can dismutate peroxide to oxygen and water (catalase activity)[51]. The present level of understanding[2] indicates only minor differences between the binuclear copper sites in arthropod and mollusc hemocyanin[52] and tyrosinase[53]. Thus the following discussion is appropriate for all of these proteins. (Spectroscopic figures in this section are of the mollusc *Busycon canaliculatum* except where otherwise noted.)

Detailed study of the coupled binuclear copper proteins has led to a spectroscopically effective picture[2] of this active site. Adequately resolved x-ray crystallographic structural information does not yet exist for any coupled binuclear copper protein; however, two groups have suitable crystals and preliminary results have been reported[54, 55]. The colorless deoxy protein form contains a binuclear cuprous ($d^{10}$) site, indicated by the lack of ligand field transitions in the optical spectrum and absence of an EPR signal. Resonance Raman studies have shown[17, 18, 19] that oxygen binds to this site as peroxide (see Introduction and Fig. 6), producing the blue oxy protein form. This two-electron reduction of dioxygen requires a two-electron oxidation of the binuclear copper site. Therefore, spectral features similar to those of normal cupric ($d^9$) sites might be expected. The oxy binuclear copper site, however, is EPR-non-detectable and, in fact, is diamagnetic to the most sensitive superconducting quantum interference detector (SQUID) magnetometers[56, 57]. Even more striking (Fig. 20) is that, in contrast to the weak low-energy ligand field bands of normal copper, two intense absorption spectral features dominate the

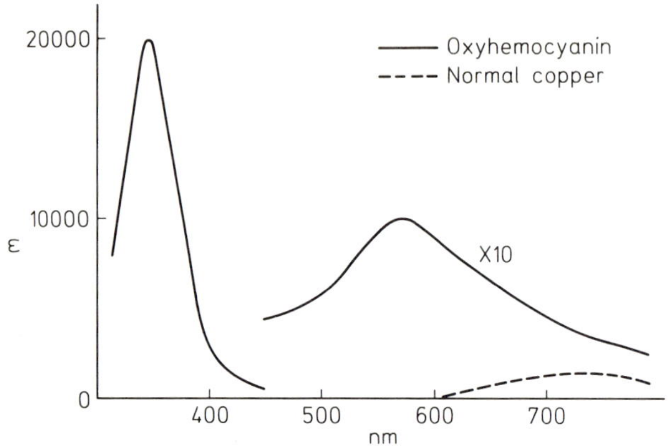


**Fig. 20.** Optical absorption spectrum of oxyhemocyanin (pH = 6.3, room temperature) and normal tetragonal copper(II)

visible-UV region: a band at about 600 nm ($\varepsilon \approx 1000\ M^{-1}\ cm^{-1}$) and another at about 350 nm ($\varepsilon \approx 20\,000\ M^{-1}\ cm^{-1}$). Elucidation of these unique spectral features has been an interestingly complicated problem due to the binuclear nature of the site.

Binuclear cupric sites would be expected to have magnetic ground state and optical spectral features significantly perturbed due to interaction of the $d^9$ metal ions. In the limit of strong interaction, direct overlap of the copper orbitals would lead to metal-metal bonding (as described by molecular orbitals) and a completely spin-paired diamagnetic ground state (Fig. 21, left). Optical transitions with energies on the order of 20 000 $cm^{-1}$ from the highest occupied molecular orbital [$\Psi_g = (1/\sqrt{2})(d_{x^2-y^2A} + d_{x^2-y^2B})$] to low excited state orbitals [$\Psi_e = (1/\sqrt{2})(d_{x^2-y^2A} - d_{x^2-y^2B})$] in metal-metal bonded complexes are allowed and can dominate the absorption spectrum. There are, however, no clearly defined cases of metal-metal bonding in binuclear cupric complexes. An indirect weaker interaction is found when bridging ligands form a superexchange pathway leading to a valence-bond type coupling of the unpaired copper spin states. When two $S = 1/2$ ions couple, they form a non-degenerate spin singlet ($S = 0$; $M_S = 0$) and a three-fold degenerate spin triplet ($S = 1$; $M_S = -1, 0, 1$); the energy difference between these states ($2J$) depends on the strength of this interaction. If the singlet is lowest in energy (antiferromagnetic coupling), a diamagnetic ground state results (Fig. 21, middle). The triplet, however, can in this case be thermally accessible at room temperature ($\sim 200\ cm^{-1}$), contributing to the paramagnetism and the EPR spectrum. Temperature dependence of these signals can thus be used to determine J. If antiferromagnetic coupling is responsible for the diamagnetism of oxyhemocyanin, then the inability to observe temperature-dependent magnetic susceptibility requires $-2J > 550\ cm^{-1}$ (*vide supra*)[56, 57]. The optical absorption spectrum is less perturbed at this weak interaction level but should still be affected by antiferromagnetic coupling. In particular, the intensity of d-d transitions could be affected, CT transitions associated with the bridging ligand would shift or split and the possibility could exist for simultaneous pair excitations[58, 59] where one photon at twice the d-d transition energy weakly excites ligand field transitions on both metal centers. These effects, however, are not well defined for small molecule spectral analogues. Poor superexchange orbital overlap or the lack of bridging ligands can lead to the

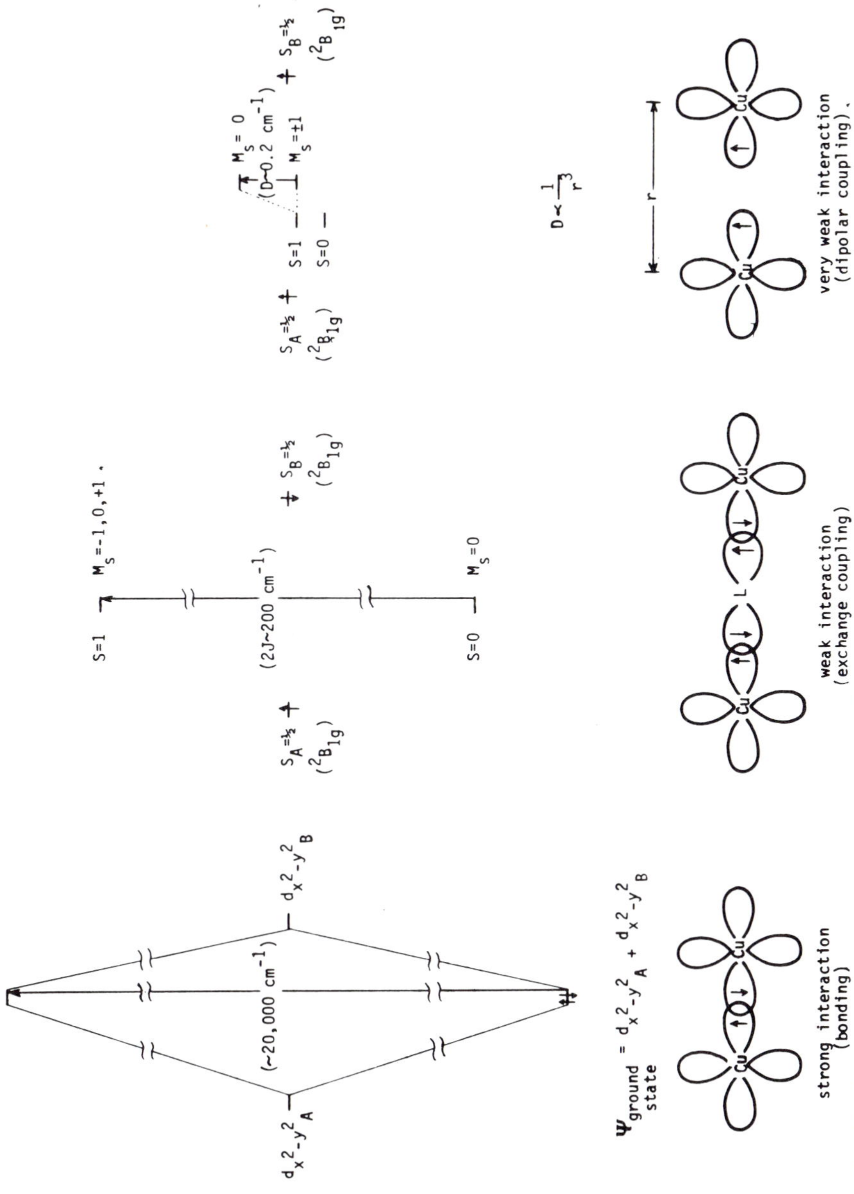


**Fig. 21.** Representation of three degrees of Cu(II)-Cu(II) interaction with associated electronic energy levels (see text for explanation)

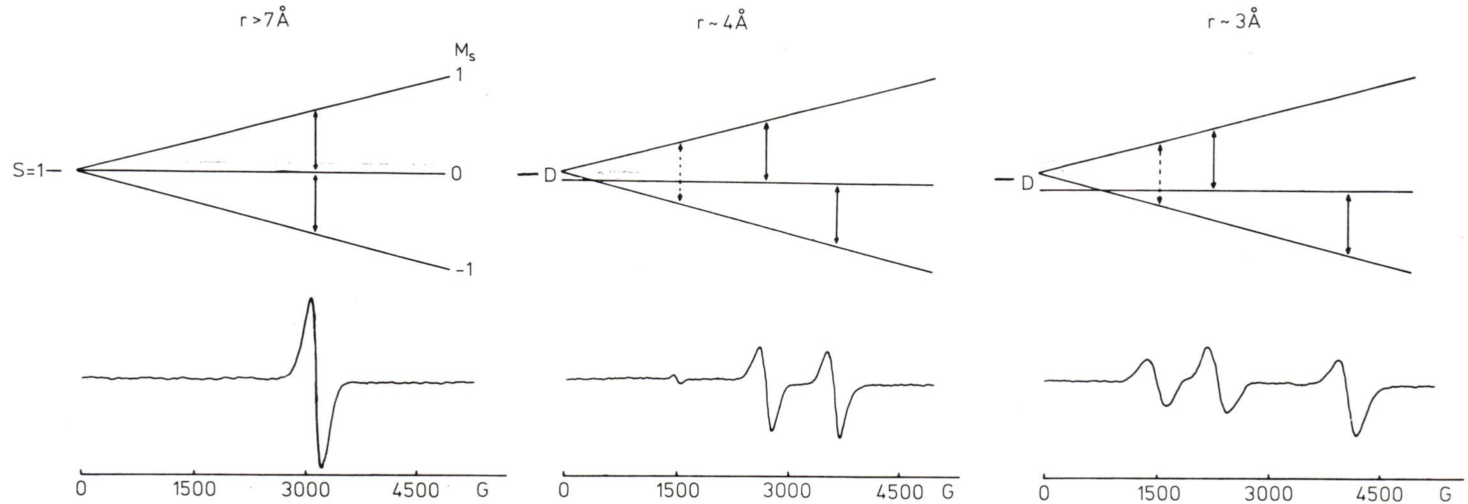


**Fig. 22.** Dipolar-induced zero field splitting ($D$) of the EPR signals of a dipolar coupled binuclear cupric spin triplet at three metal-metal separations ($r$); solid lines are $\Delta M_s = \pm 1$ transitions and broken lines are $\Delta M_s = \pm 2$ transition; anisotropy and hyperfine contributions have been neglected

limit ($2J \rightarrow 0$) of very weak interaction (Fig. 21, right). Here, only the EPR spectrum[60] is perturbed through dipolar coupling of the spins (Fig. 22). This interaction, which depends on the separation (r) of the cupric ions, produces a zero field splitting [$D_{dipolar} \approx -(g_{\|}^2 + g_{\perp}^2/2)(\beta^2/r^3)$] of the spin triplet, indicated by broadening and eventual splitting of the $\Delta M_S = \pm 1$ transitions in the EPR spectrum as the Cu–Cu separation decreases. A $\Delta M_S = \pm 2$ transition at half field also becomes weakly allowed, reaching maximum intensity when $D \approx h\upsilon/2$ (~.15 $cm^{-1}$ for X-band, corresponding to $r \approx 3$ Å).

A series of hemocyanin and tyrosinase active site derivatives (Fig. 23) can be prepared[61–66], allowing systematic variation of the binuclear copper active site and chemical perturbation for spectral studies. In the simplest derivative, met-apo, one copper has been removed and the remaining copper oxidized to the spectroscopically accessible Cu(II). Next in complexity is a mixed-valent binuclear copper site. The Cu(II), in this half-met derivative, exhibits open-shell $d^9$ spectroscopic features and the Cu(I), though spectroscopically inaccessible, can still be studied by comparison to the met-apo derivative. Two derivatives have formally binuclear cupric sites: met, which is EPR-non-detectable, and dimer, which exhibits an intense broad EPR signal. Spectroscopic study of these derivatives has led to the present picture of the coupled binuclear copper protein active site shown at the bottom of Fig. 23.

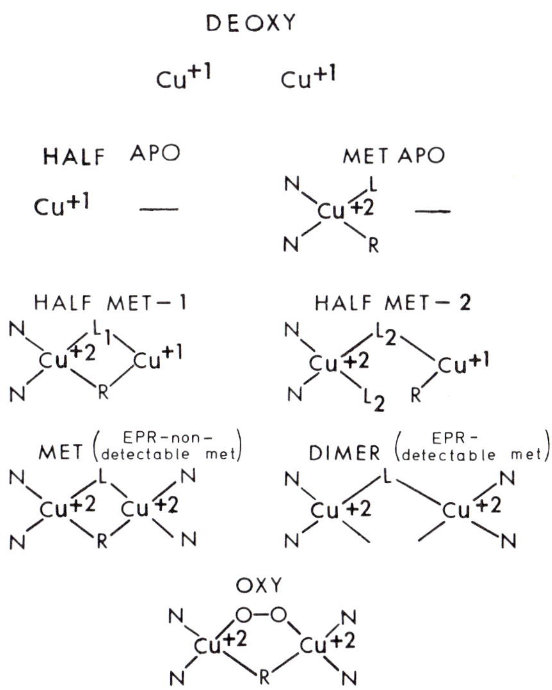


**Fig. 23.** Representations of the active site derivatives of hemocyanin and tyrosinase (nomenclature as suggested at the European Molecular Biology Meeting on "Comparative Study and Recent Knowledge on Quaternary Structure and Active Sites of Oxygen Carriers and Related Proteins", Tours, France, August 20–24, 1979)

## B. Exogenous Bridging Ligands

Chemical and spectroscopic comparison of the met-apo and half-met derivatives demonstrates that exogenous ligands bridge the two coppers. The EPR signals of Fig. 24 show that exogenous ligands, such as $N_3^-$, bind to the Cu(II) of met-apo, yet dialysis rapidly removes the ligand. This indicates labile ligand-substitution chemistry as found for normal tetragonal Cu(II). The spectral features are also consistent with a tetragonal Cu(II) site (d-d $\approx$ 750 nm, $\varepsilon \approx 250\ M^{-1}\,cm^{-1}$; $N_3^- \rightarrow$ Cu(II) CT $\approx$ 420 nm, $\varepsilon \approx 2500\ M^{-1}\,cm^{-1}$, $g_{\parallel} = 2.314$, $g_{\perp} = 2.083$, $A_{\parallel} = 131 \times 10^{-4}\ cm^{-1}$)[62,64]. In strong contrast, the half-met derivative binds azide quite tightly – a new EPR signal remains even after extensive dialysis. This chemical evidence suggested that the Cu(I) as well as the Cu(II) (azide-to-Cu(II) charge transfer transitions are also found for half-met $N_3^-$) is involved in binding exogenous ligands. Further, CO can reversibly bind to the half-met Cu(I), inducing changes in the EPR signal of the Cu(II) and labilizing the bound exogenous ligand. Clearly both metals are involved in binding exogenous ligands at the site.

Figure 25 presents quite interesting spectral data[64] which also require that exogenous ligands bridge the two copper ions. A series of half-met derivatives can be prepared with various anions tightly bound at the active site. Those anions that are more capable of electron delocalization exhibit more complicated EPR signals in the $g_{\parallel}$ region (more than four copper parallel hyperfine lines) indicating coupling to the second copper. Half-met $N_3^-$ even begins to show a curious isotropic signal, indicative of rapid electron exchange between the two coppers.

The optical spectra of these half-met forms also show a new feature that decreases in energy and increases in intensity as the electron delocalizing ability of the exogenous anion increases. This is assigned as an intervalence transfer (IT) transition which is a special type of charge transfer process corresponding to the optical excitation of an electron from $Cu_A$(I) to $Cu_B$(II). In the limit of no interaction between the coppers, defined as class I[67] mixed valence, the ground state is described as $\Psi_g = Cu_A(I)Cu_B(II)$ and the IT excited state as $\Psi_{IT} = Cu_A(II)Cu_B(I)$. When the coppers interact, the ground state becomes $\Psi'_g = (1 - \alpha^2)^{1/2}\Psi_g + \alpha\Psi_{IT}$. For $\alpha = \sqrt{.5}$ the site is completely delocalized, i.e., Cu(1.5)Cu(1.5) and is defined as class III[67] mixed valence. When $0 < \alpha < \sqrt{.5}$, the system is class II[67] mixed valence and the spectral features are the sum of those for Cu(II) and Cu(I) with additional new features associated with delocalization. Class II assignment is appropriate for the half-met derivatives where coupling to the Cu(I) is demonstrated by appearance of more than four copper hyperfine lines in the EPR spectrum and the IT transition. Intensities of IT transitions are proportional to $\alpha^2$ and must be mediated by the bridging ligand, the pathway for electron transfer. Therefore, since intensity of the half-met IT transition and degree of EPR hyperfine coupling to the second copper correlate with covalency interactions of exogenous ligands, it is demonstrated that the exogenous ligand must bridge the two coppers at this site.

LEFE data for the half-met derivatives can, in principle, probe mixing of the IT excited state into the ground-state wavefunction (see Introduction). Half-met $NO_2^-$, which exhibits no CT or IT features and a normal cupric EPR signal, shows an average $\sigma$ (Fig. 26) which is large compared to tetragonal cupric complexes (Fig. 2C), indicating a significant non-centrosymmetric distortion. The orientation dependence of this effect indicates that it is largest with $E \parallel g_z$. These observations indicate a $\sim C_{4v}$ site symmetry for the Cu(II). Preliminary LEFE data[68] (Fig. 26) for a series of half-met derivatives

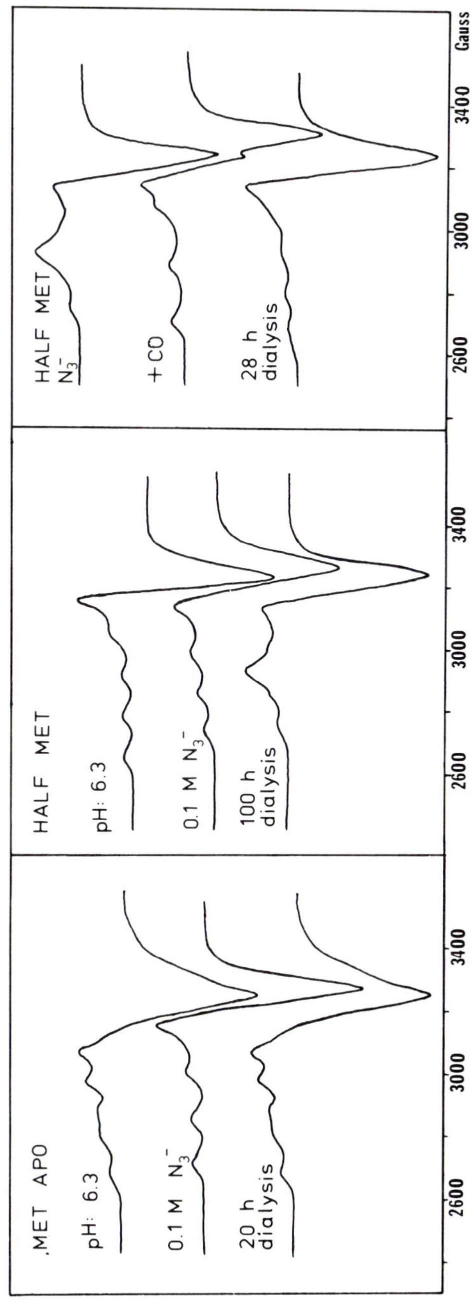


**Fig. 24.** Met-apo and half-met hemocyanin EPR spectra (77 K, $v$ = 9.1 GHz, pH = 6.3) perturbed by anion binding and dialysis (adapted from Ref. 64)

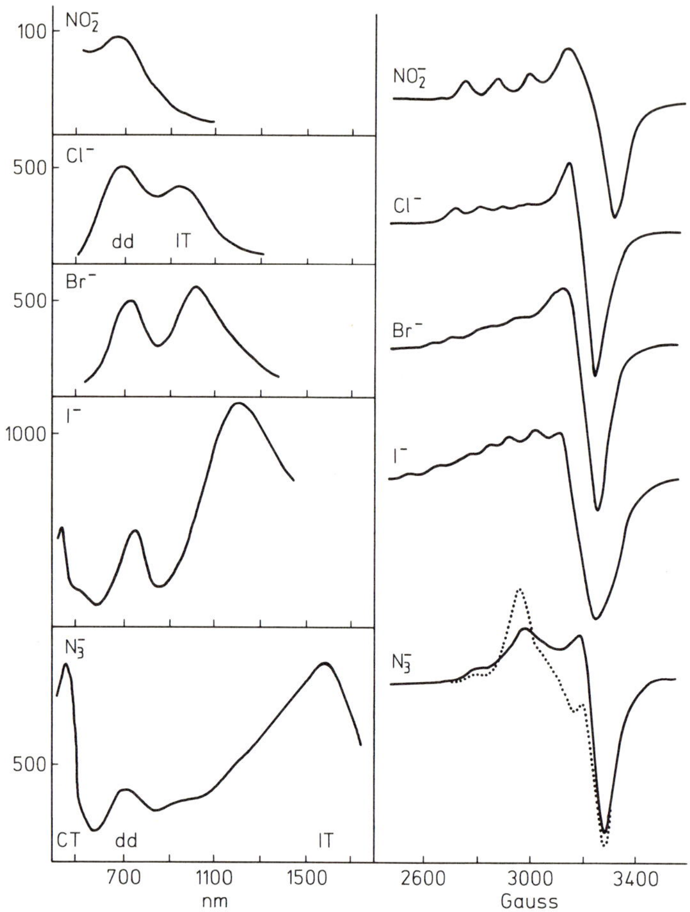


**Fig. 25.** Absorption spectra (~ 15 K, pH = 6.3 in 1 : 1 sucrose glass) and EPR spectra (77 K, $v$ = 9.1 GHz) of the half-met hemocyanin series; dotted line for $N_3^-$ is 7 K spectrum (from Ref. 52)

possessing low-energy CT and IT transitions indicate that displacement of the maximum shift away from $g_z$ correlates with the exogenous ligand-to-Cu(II) CT transition, (oriented approximately in the equatorial plane) rather than the direct IT transition. This supports equatorially polarized charge transfer mixing into the ground-state wavefunction of the half-met site.

Studies of electronic analogues for the half-met derivatives could lead to correlation of electronic features (EPR and IT transition) of this mixed valence site to specific bridging geometric features. Mixed valence spectral features could also be used to probe the normally spectroscopically inaccessible Cu(I) energy levels. Electrochemical studies of binuclear copper complexes[69, 70] suggest that ligand flexibility and distortion are important factors in stabilizing this Cu(II)Cu(I) site. A number of mixed valence binuclear copper complexes have been made, some of which show distinct class II spectral features[71, 72]. However, meaningful correlation of structural and spectral features is

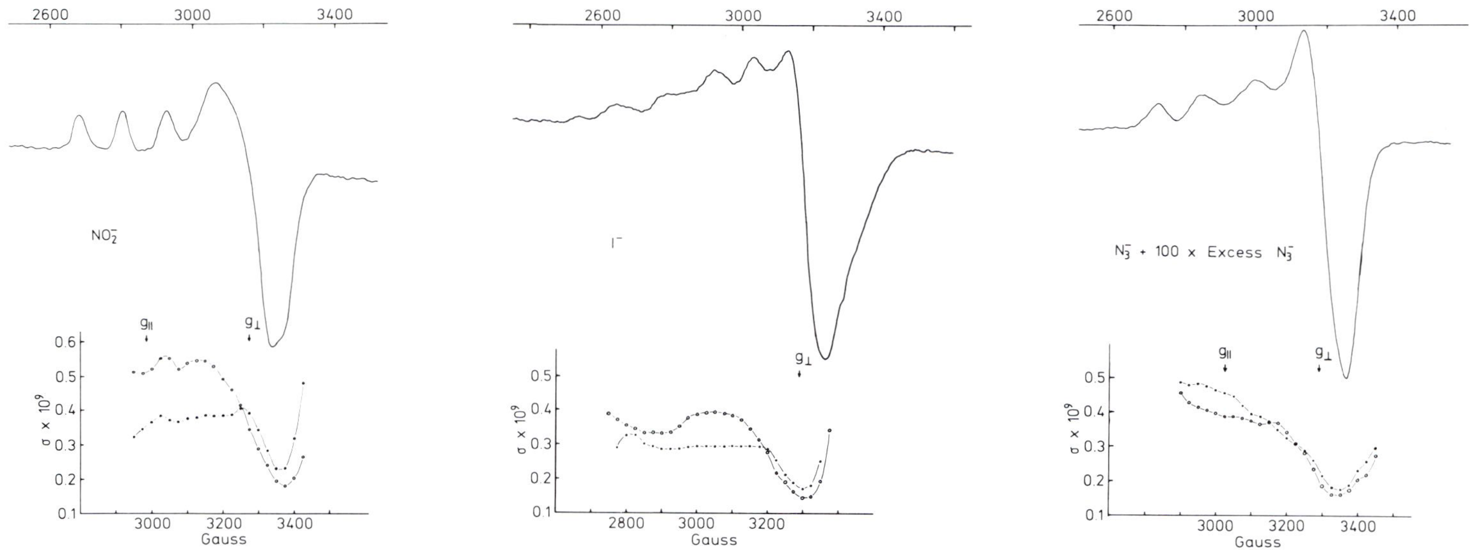


**Fig. 26.** Half-met hemocyanin EPR spectra (~ 7 K, $\nu$ = 9.1 GHz, pH = 6.3) and associated LEFE signals[68] (4.2 K; pH = 6.3); $NO_2^-$: $\nu$ = 9.61 GHz; $I^-$: $\nu$ = 9.45 GHz; $N_3^-$ with excess $N_3^-$: $\nu$ = 9.50 GHz; ○ ≡ E‖H and ● ≡ E⊥H; note that EPR and LEFE spectra are positioned such that $g_\parallel$ and $g_\perp$ of each are aligned

precluded by the inability to produce systematic variation in site structure among those complexes presently well characterized through x-ray crystallography.

## C. Endogenous Ligand Bridge

Chemical and spectral studies have indicated that in addition to bridging exogenous ligands, there is an endogenous protein bridge at the coupled binuclear copper site. Evidence for this is seen with the half-met derivative and the binuclear cupric derivatives met and dimer.

Exogenous anions can be divided into two groups based on their half-met binding characteristics (Fig. 27). Group 1 ligands ($NO_2^-$, $F^-$, $Cl^-$, $Br^-$, $I^-$, $Ac^-$) tightly bind at the site with a Cu–Cu distance of less than 4 Å and show no change in EPR signal when present in excess. Group 2 ligands ($N_3^-$, $SCN^-$, $CN^-$), however, can bind with a Cu–Cu separation of greater than 5 Å and show major changes in their EPR signal when present in excess. These changes indicate that a second group 2 ligand binds at the site.

Optical absorption and resonance Raman studies of half-met $N_3^-$ shown in Fig. 28 have indicated the nature of the group 2 ligand binding. Half-met $N_3^-$ shows a quite low energy $N_3^- \rightarrow$ Cu(II) CT transition at ~500 nm. In this case the low energy (compared to Fig. 5) is due to bridging steric effects on the $N_3^- \rightarrow$ Cu(II) bonding. Laser excitation into this band reveals a resonance-enhanced Raman spectrum with metal-ligand (397 $cm^{-1}$) and intra-ligand (2039 $cm^{-1}$) stretching frequencies indicative of a bridging $N_3^-$. Half-met $N_3^-$ with excess azide shows an additional CT feature at ~400 nm, an energy more consistent with that of normal tetragonal Cu(II)-$N_3^-$ complexes. Excitation into the 500 nm band of the sample with excess azide shows stretching frequencies identical to those of the sample with no excess (intense fluorescence at higher excitation energies

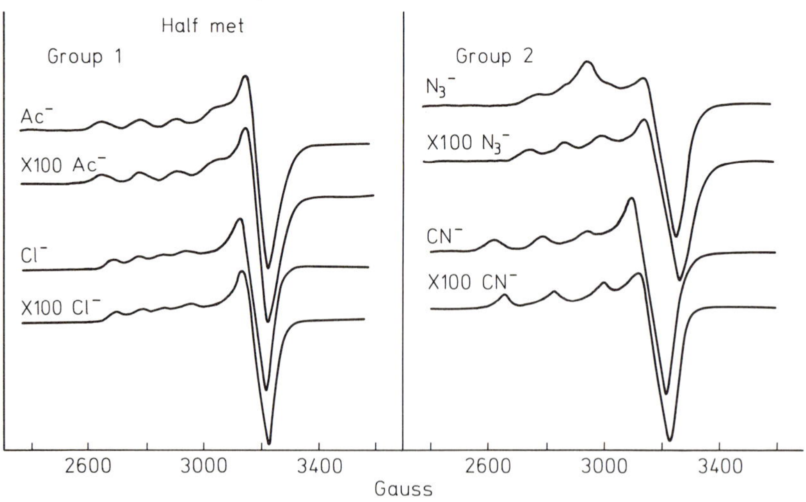


**Fig. 27.** EPR spectra (77 K, $v = 9.1$ GHz; pH = 6.3) of group 1 and group 2 half-met hemocyanin derivatives with and without 100-fold excess exogenous ligand (adapted from Ref. 64)

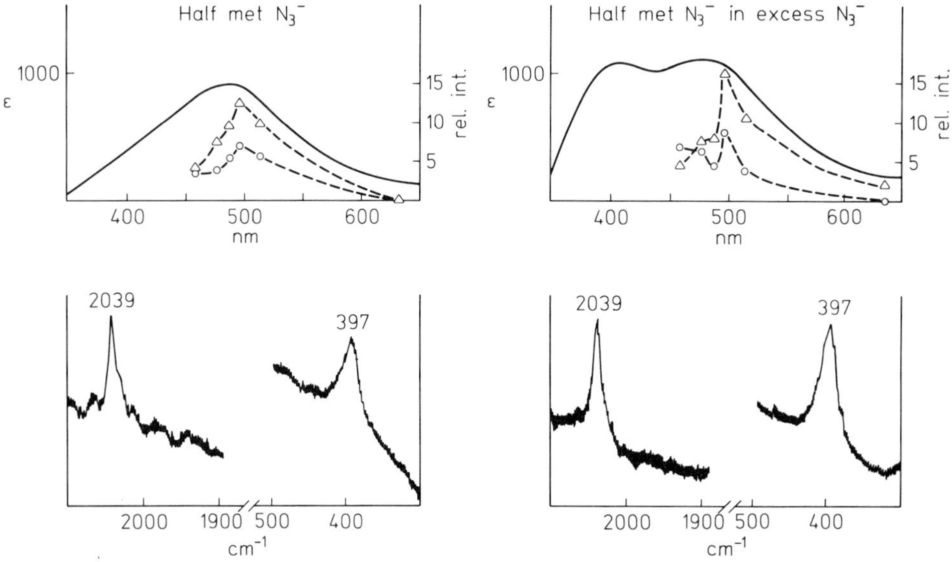


**Fig. 28.** Optical absorption (room temperature difference spectra) (*upper*) and resonance Raman spectra (room temperature) (*lower*) of half-met $N_3^-$ and half-met $N_3^-$ with 100-fold excess $N_3^-$; excitation profiles ($\Delta$ = 2039 $cm^{-1}$, O = 397 $cm^{-1}$) correspond to right side scale (from Ref. 2)

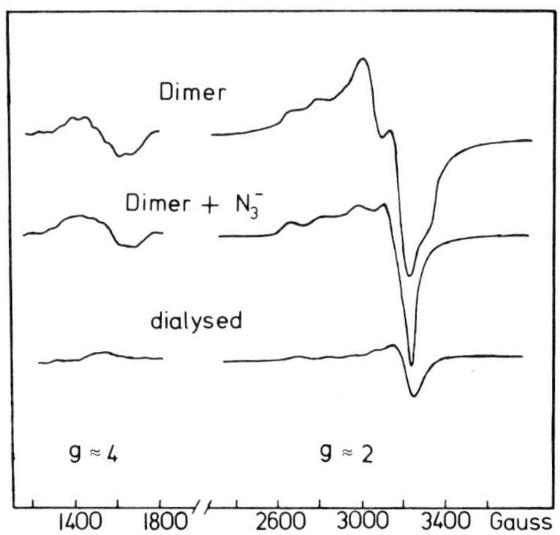


**Fig. 29.** EPR spectra (77 K; $v$ = 9.1 GHz; pH = 5.7) of dimer, dimer with 100-fold excess $N_3^-$, and after 24-h dialysis; 10 mW power in g ≈ 2 region and 100 mW power in g ≈ 4 region (from Ref. 52)

precludes probing the ~ 400 nm band). This demonstrates that the second ligand which binds in the excess form does not perturb the bridging ligand.

Dimer, prepared from half-met $NO_2^-$ by reaction with $N_2O_4$[73], exhibits a broad EPR signal[74] (Fig. 29) in the g ≈ 2 region and a weak signal at g ≈ 4. The temperature dependence of these signals obeys Curie behavior down to 14 K, demonstrating negligible exchange interaction ($|2J| < 5$ $cm^{-1}$). This type of signal thus is associated with zero field splitting of the triplet state of two dipolar coupled copper(II)'s (see Fig. 22). Simulation of this EPR signal[75] with computer programs written by Pilbrow[60] indicates a Cu–Cu distance for the two tetragonal Cu(II)'s ($g_{\|} = 2.39$; $g_{\perp} = 2.07$; $A_{\|} = 170 \times 10^{-4}$ $cm^{-1}$) of ~6 Å[76]. Therefore, dimer has a group 2 ligand and is analogous to the group 2 ligand half-met form. Ligand substitution with other group 2 anions (such as azide) preserves this type of signal (a larger zero field splitting indicates a somewhat shorter Cu–Cu distance). Dialysis, however, causes this EPR signal to disappear, resulting in weak signals and indicating that dimer converts to met when the group 2 ligand is removed[65].

Met is the resting form of tyrosinase; it is prepared in mollusc hemocyanin by ligand displacement of peroxide from oxy[77] while in arthropods, two-electron oxidation of deoxy with peroxide leads to met[78]. As indicated above, weak EPR signals (*vide infra*) are seen[79] for met; however, they are due[63, 80] to a small fraction of sites that have become disrupted. This is clearly demonstrated by regenerating mollusc met to oxy with peroxide, as shown in Fig. 30. More than 90 percent of the sites can be regenerated, yet the weak signals remain unchanged. Hence, met is a formally binuclear cupric derivative which, like oxy, is EPR-non-detectable.

Ligand field transitions provide the most appropriate spectral comparison for the two formally binuclear cupric derivatives. As shown in Fig. 31, the d-d transitions of met and dimer are very similar and indicative of two tetragonal Cu(II)'s. The similarity of the two spectra reflects similar electronic structure in the two sites. Lack of an EPR signal, therefore, results from only a weak interaction on the energy levels of the copper(II)'s. The appropriate weak interaction is an antiferromagnetic coupling of the spins which requires a superexchange pathway, as mentioned previously. Therefore, an endogenous protein bridge is necessary.

The parallel analogies between group 1 and group 2 half-mets and between met and dimer provide clear evidence for the endogenous protein bridge. Group 1 exogenous ligands with a Cu–Cu separation of less than 4 Å maintain the endogenous bridge in half-

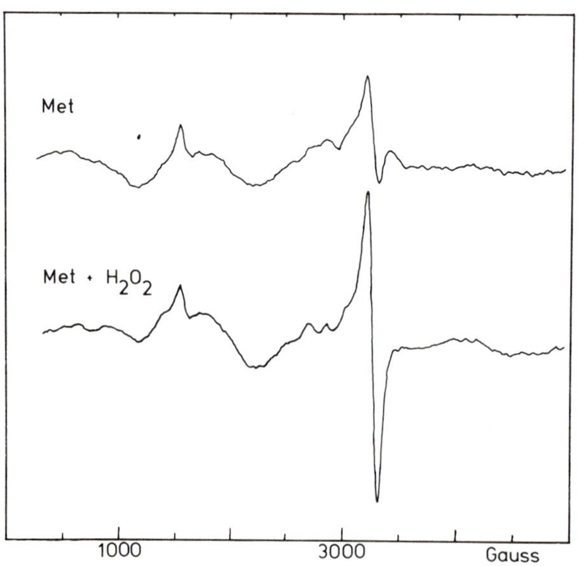


**Fig. 30.** EPR spectra (7 K; $\upsilon$ = 9.2 GHz, pH = 5.0) of met and met treated with 10-fold excess $H_2O_2$ (from Ref. 2)

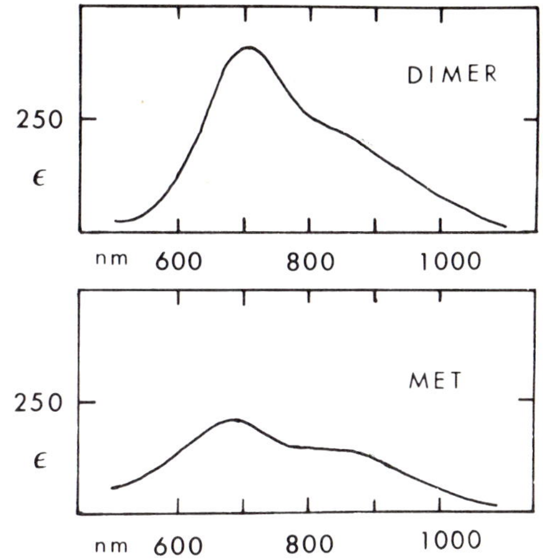


**Fig. 31.** Optical absorption spectra (~ 15 K) of dimer (pH = 5.0) and met (pH = 6.3) in the ligand field region (from Ref. 2)

met. Similarly, the endogenous bridge provides the superexchange pathway in met (Cu–Cu also less than 4 Å from EXAFS studies[81)]) leading to coupled spins and lack of an EPR signal. Group 2 ligands force a separation greater than 5 Å in half-met, opening up a second exchangeable position at the Cu(II) when the endogenous bridge is broken. Group 2 ligands also maintain this larger separation in dimer, breaking the endogenous superexchange pathway and leading to EPR-detectable dipolar coupled Cu(II)'s.

Detailed study of the weak EPR signals associated with met hemocyanin[80)] has provided some insight into the nature of the endogenous protein bridge. These broad (0 to 5000 Gauss) signals (Fig. 32) are reversibly formed at low pH when the endogenous bridge in a heterogeneous fraction of the sites becomes protonated and replaced by $Ac^-$. Since the temperature dependence of the EPR signal intensity indicates that these sites contain dipolar coupled Cu(II)'s, computer simulations (*vide supra*) show that exogenous ligands modify the distance between the uncoupled metals (group 1: $Br^- \sim 3$ Å; group 2: $N_3^- \sim 5$ Å). However, these exogenous halide and pseudo-halide ligands do not mediate significant exchange coupling between the copper(II)'s in contrast to the endogenous bridge.

Since EXAFS studies[82, 83)] have indicated that sulfur is not strongly interacting with the protein site, this leaves nitrogen and oxygen donor ligands as possible endogenous bridges. Model studies (see Table 4) have indicated that of possible biological ligands, only phenoxide, alkoxide or hydroxide bridges are capable of mediating the large exchange interaction found in the coupled binuclear copper protein site. This coupling is strongly dependent on the Cu–O–Cu angle (Goodenough-Kanamori rules[85)]), where angles approaching 180° are required for effective superexchange interaction and large antiferromagnetic splittings. Experimentally it is found for planar tetragonal $Cu_2^{+2}(\mu\text{-}O^-)_2$ complexes that angles greater than 105° provide $-2J > 500\ cm^{-1}$[84a)]. The intrinsic $pK_a$'s of these possible endogenous bridges, phenolate ($pK_a = 10.1$), water ($pK_a = 15.7$), serine or threonine ($pK_a > 16$), indicate that tyrosine is more likely to be stabilized to an apparent $pK_a$ well below 5.0.

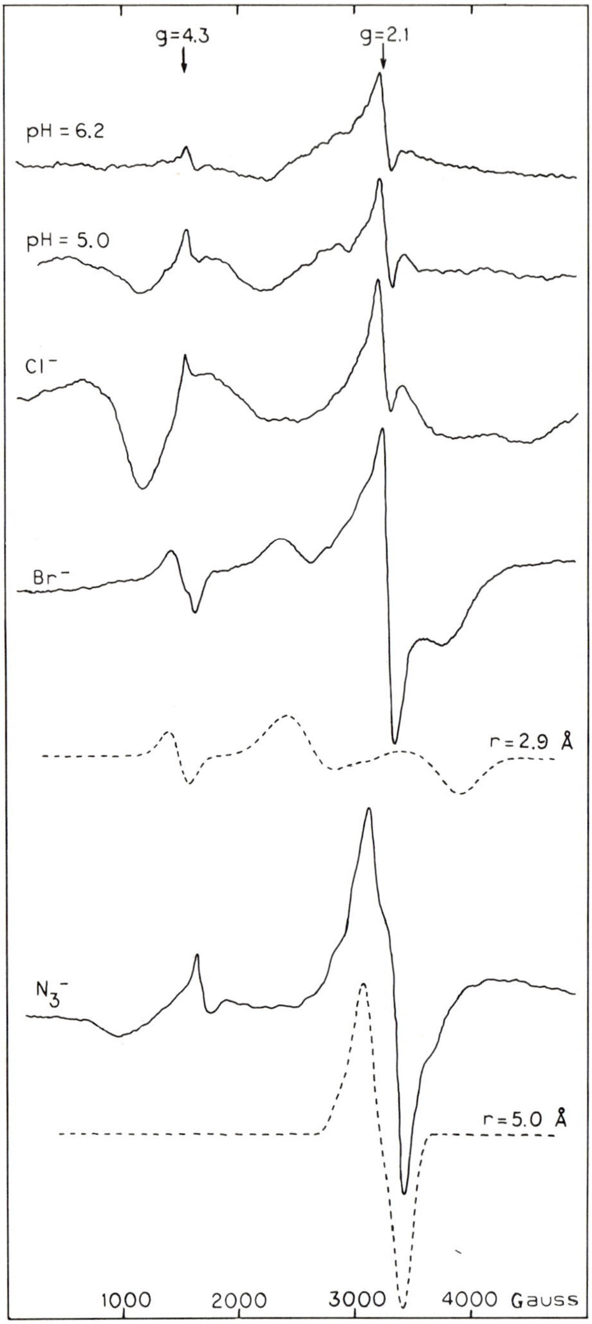


**Fig. 32.** EPR spectra[68] of met hemocyanin (1–2 mM protein concentration; ~7 K; $v$ = 9.2 GHz, 10 mW power) under the indicated conditions; anions in > 5-fold excess at pH = 5.0, 0.1 M $Ac^-$ buffer; sharp signal at g = 4.3 is ubiquitous "rhombic iron" impurity; mononuclear $Cu^{+2}$ signal at g = 2.1 is due to sites irreversibly damaged by large anion excess; *dotted lines* are simulations with Cu–Cu distance (r) indicated

A charge transfer transition is found at 425 nm in both met and oxy (*vide infra*) and is characteristic of the coupled binuclear copper site. Elimination of sulfur ligation leaves the most reasonable assignment of this feature as a phenolate-to-copper(II) CT transition (see Table 4). This indicates that tyrosine is bound to the binuclear cupric site and the most likely candidate for the endogenous bridge.

**Table 4.** Properties of copper(II) complexes with phenolate, alkoxide and hydroxide ligands

A. Optical features and exchange coupling for structurally defined binuclear copper(II) complexes

| Bridging ligand type | Complex | $Cu^{+2}$ Geometry | $-2J$ ($cm^{-1}$) | CT (nm)[a] | d-d($\lambda_{max}$) (nm)[a] | Ref. |
|---|---|---|---|---|---|---|
| $\mu$-$^{-}$OH | $[Cu_2(OH)(ClO_4)A]^{+2}$ (binucleating N donor macrocycle; equatorial $\mu$-OH; axial $\mu$-$ClO_4$) | $D_{2d}$ distorted square planar | 500 | 330(1800) | 637(240) | 84b |
| $\mu$-$^{-}$OH | [Cu(NNN'N' tetramethyl ethylene) $(OH)]_2^{+2}$ | square planar | 360 | 355 R | 565 R | 84c, d |
| $\mu$-$^{-}$O-alkyl | $Cu_2C_{20}H_{26}N_2O_4F_{12}$ (two equatorial bonding alkoxide, imine, $\mu$-alkoxide ligands) | square planar | ≥600 | 356(4170)D | 590(163)D | 84e, f |
| $\mu$-$^{-}$O-alkyl | $Cu_2C_{20}H_{42}N_4O_{12}Cl_2$ (two equatorial bonding amine, imine, μ-alkoxide and axial alkoxide ligands) | distorted square pyramidal | small | 379 R | 625 R | 84g |
| $\mu$-$^{-}$O-phenyl | $Cu_2LBr_2(ClO_4)MeOH$ (tetra amine, $\mu$-phenoxide ligand with $\mu$-Br; axial Br and MeOH) | square pyramidal and distorted trigonal bipyramidal | 139 | 425–450 R | 650 R | 84h |
| $\mu$-$^{-}$O-phenyl | dichloro bis(N-ethylsalicylaldimino) dicopper(II) | $D_{2d}$ distorted square planar | 480 | [b] | 621, 794 N | 84i, j |

B. Optical features of structurally defined mononuclear copper(II) phenolate complexes

| Complex | $Cu^{+2}$ Geometry | CT(nm)[a] | d-d($\lambda_{max}$)(nm)[a] | Ref. |
|---|---|---|---|---|
| Cu(2,4,6-trichlorophenolato)$_2$ (imidazole)$_2$ | distorted octahedral equatorial phenolate | 430(1200)A | 710(150)A | 86a, b |
| Cu(2,4,6-trichlorophenolato)$_2$ (tmen) | $D_{2d}$ distorted square planar | 487(2500)B | 746 sh, R | 86c, d |
| Cu(2-methoxy-4-nitrato-phenolato)$_2$ (tmen) | distorted octahedral equatorial phenolate | 362, 425 sh, 472 R | 541, 791 sh, R | 86e, d |
| Cu(4-formyl-2-methoxy-phenolato)$_2$ (pyridine)$_2$ | distorted octahedral equatorial phenolate | 415 sh, 476 R | 610, 800 R | 86f, d |

**Table 4** (continued)

C. Optical features of non-crystallographically defined copper(II) phenolate complexes

| Complex | CT(nm)[a] | d-d($\lambda_{max}$)(nm)[a] | Ref. |
|---|---|---|---|
| Cu($aa'$-ethylenediimino-di-$o$-cresol | 328(1000), 400(1000)M | 588(260)M | 86g, 12a |
| Cu(II)(d$^{\ell}$-$o$-tyrosine)$_2$ (pH = 10.5)[c] | 393(590) | 630(88) | 86h |
| Cu(II) + transferrin (pH = 8.5)[d] | 430(1890) | 710(235) | 86b |
| Cu(II) + lactoferrin (pH = 7.6)[d] | 438(2400) | 677(310) | 86i |

[a] $\varepsilon$/Cu$^{+2}$(M$^{-1}$ cm$^{-1}$) indicated in parentheses for aqueous solutions except where noted; R = reflection spectrum; N = Nujol mull; D = dichloromethane; A = acetone; B = benzene; M = methanol + 2 eq. NaOH; sh = shoulder

[b] Phenolate → Cu(II) CT transitions are not easily distinguished for salicylaldimino complexes as ligand-centered $\pi \rightarrow \pi^*$ transitions can be found as low as 400 nm[84k)]; tyrosine $\pi \rightarrow \pi^*$ transitions occur at 275 nm

[c] Even though the phenol is deprotonated under these conditions, $m$- and $p$-tyrosine show no phenolate → Cu(II) CT transitions supposedly due to steric considerations[86j)]; serine and threonine alkoxide bonding interactions with Cu$^{+2}$ are not known[86k)]

[d] Resonance Raman studies[86b, l)] of these systems (excitation into the CT feature) show enhanced characteristic phenol vibrations

## D. Oxy Site

With this understanding of the various coupled binuclear copper derivatives, and an explanation of the ground-state magnetic properties, the unique optical spectral features of oxyhemocyanin can be interpreted. Comparing met and oxy (Fig. 33) allows identification of those features directly associated with the bound peroxide. The two dominant optical features ($\lambda$ = 345 nm, $\varepsilon \approx$ 20000 M$^{-1}$ cm$^{-1}$; $\lambda$ = 580 nm, $\varepsilon \approx$ 1000 M$^{-1}$ cm$^{-1}$) are not present in met, demonstrating that they can be assigned as peroxide-to-copper(II) charge transfer transitions. Resonance Raman excitation into these bands (see Introduction) reveals peroxide stretching modes, further supporting this argument. The ligand field bands are similar (oxy $\lambda_{max} \approx$ 700 nm, $\varepsilon \approx$ 200 M$^{-1}$ cm$^{-1}$; met $\lambda_{max} \approx$ 690 nm, $\varepsilon \approx$ 200 M$^{-1}$ cm$^{-1}$), reflecting a common tetragonal geometry for the two Cu(II)'s. As mentioned above, a feature at 425 nm is also common and most likely due to a phenolate-to-Cu(II) CT transition, probably from the endogenous bridge. Of particular importance in the CD spectrum of oxy (Fig. 33) is an additional $O_2^{=} \rightarrow$ Cu(II) CT transition (480 nm) which is magnetic dipole allowed (large Kuhn anisotropy factor, $\gamma$ = 0.02; strong in CD but not in the absorption spectrum) and which must be interpreted for a complete assignment of the electronic structure. The observation of at least three $O_2^{=} \rightarrow$ Cu(II) CT transitions requires that peroxide bridge the two coppers. Only two CT transitions are expected from peroxide bound to one tetragonal copper(II) (see Fig. 34). The highest-energy degenerate p$\pi$ orbitals in peroxide split into two nondegenerate orbitals on bind-

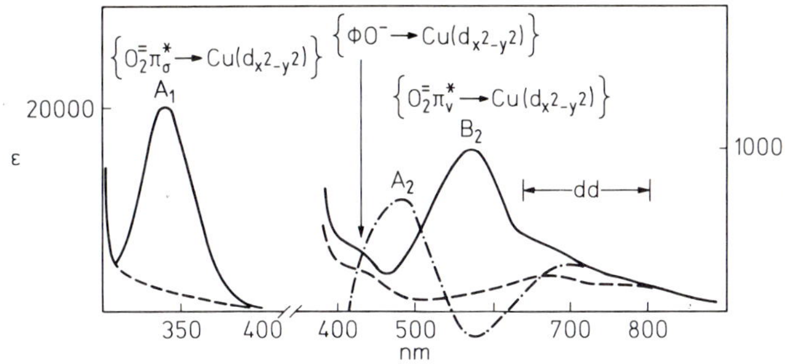


**Fig. 33.** Optical absorption spectra (~ 15 K) of oxy (—) and met (---) hemocyanin and CD spectrum (room temperature) of oxy (-·-·-) with assignment of features; all samples at pH = 5.0 (from Ref. 66)

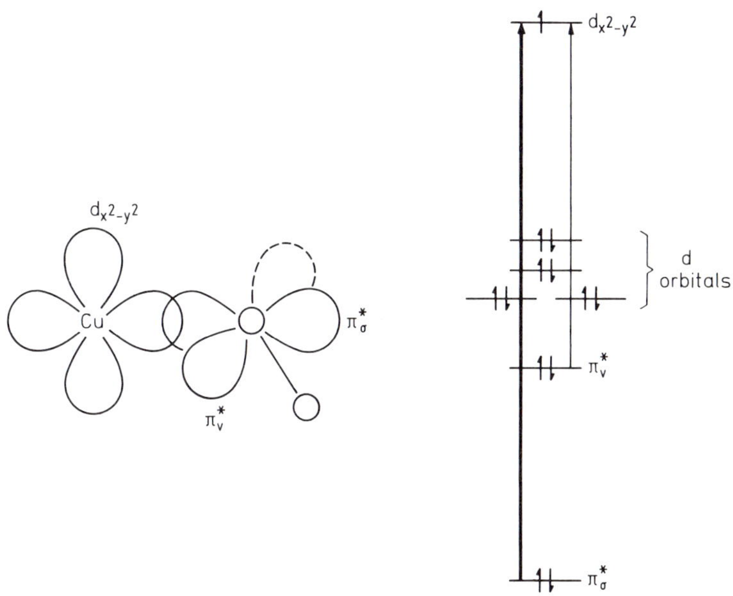


**Fig. 34.** Copper d orbital and peroxide $\pi^*$ orbital splitting diagram for a copper-peroxide monomer; intensity of the charge transfer transition, based on orbital overlap considerations, is indicated by width of the *arrows*

ing. $\pi_\sigma^*$ is $\sigma$ bonding to the $d_{x^2-y^2}$ Cu orbital; a CT transition from it will be found at high energy with strong intensity due to extensive orbital overlap. $\pi_v^*$ is $\pi$ bonding and the charge transfer transition associated with it will be at lower energy and of weaker intensity. When these monomer interactions are extended to a weakly coupled binuclear geometry, the two CT transitions should each split into a symmetric and an antisymmetric transition. Group theoretical considerations allow prediction of the selection rules for CT transitions in this weakly coupled $C_{2v}$ binuclear site. The low-energy, weak-intensity

$\pi_v^* \rightarrow Cu(II)$ CT transition of the monomer splits into an electric and magnetic dipole allowed transition ($B_2$) and an exclusively magnetic dipole allowed transition ($A_2$). This is exactly what is found in the oxy coupled binuclear copper site (Fig. 33) where the optical and CD feature at ~570 nm and the CD feature at ~480 nm correspond to these two transitions. Only one component of the higher-energy $\pi_\sigma^* \rightarrow Cu(II)$ CT transition, the 350 nm optical peak, can be observed due to the intense UV protein absorption. This is assigned as the $A_1$ component[87].

The energy and intensity of these CT transitions can be correlated with the results of a simple model[66] to predict the likely bridging mode of peroxide at the site. CT transitions for two possible bridging geometries, $\mu$-dioxo and $\mu$-monooxo, can be treated as transition dipole vectors. Based on the total transition intensity, vector coupling (Fig. 35) generates the relative CT intensities ($I_A/I_B$) and the splitting of these transitions [$\Delta E(A_1 - B_1)$]:

$$\frac{I_{A_1}}{I_{B_1}} = \frac{|\vec{M}_1 + \vec{M}_2|^2}{|\vec{M}_1 - \vec{M}_2|^2} \tag{7a}$$

$$\Delta E(A_1 - B_1) = \frac{-2|\vec{M}|^2}{R^3} |3\cos^2\theta - \cos\phi| \tag{7b}$$

where $|\vec{M}|$ is the magnitude of the transition dipoles at distance R and the angles indicate orientation of the transition dipole vectors with respect to the Cu–Cu axis ($\theta$) and each other ($\phi$). For the $\mu$-monooxo geometry, this analysis predicts the lowest-energy $A_1$ component to have only a quarter of the total $\pi_\sigma^* \rightarrow Cu(II)$ CT intensity. Associating this component with the intense 350 nm band, the total CT intensity is predicted to be far too large and an unrealistic splitting of the $A_1$ and $B_1$ components is found. Alternatively, the $\mu$-dioxo geometry predicts that the lowest-energy component has most of the transition intensity. An analogous assignment of the 350 nm band as the lowest-energy $A_1$ component for this case indicates a reasonable total $\pi_\sigma^* \rightarrow Cu(II)$ CT intensity and splitting of the symmetric and antisymmetric components. Finally, the binuclear splitting of each $\pi^* \rightarrow Cu(II)$ CT transition can be theoretically removed and the resultant $\pi_\sigma^*$-$\pi_v^*$ splitting (~12000 $cm^{-1}$) of the $\mu$-dioxo geometry is consistent with results obtained for cobalt-(III) $\mu$-peroxo complexes (*vide infra*). This treatment and assignment of the charge transfer spectrum of oxy therefore strongly supports the $\mu$-dioxo bridging geometry and finally leads to the spectroscopically effective active site shown in Fig. 36.

The binuclear cobalt(III) $\mu$-peroxo compounds which have been structurally characterized[88–91] provide appropriate spectral analogues for the oxy active site. A preliminary analysis of the $Co(III)_2$ $\mu$-$O_2^=$ spectral features[92] has led to an empirical correlation between $O_2^= \rightarrow M$ CT transitions and peroxide bridging geometry. For the $\mu$-dioxo geometry, energy and intensity differences of the $\pi_v^* \rightarrow M$ and $\pi_\sigma^* \rightarrow M$ CT transitions seem to depend on the torsional angle between the two O–M bonds with maximum differences (~6000 $cm^{-1}$)[93] expected for the planar case[94]. Splitting of each CT transition into a symmetric and antisymmetric component in these binuclear sites has not yet been considered. Upon protonation of peroxide and formation of the $\mu$-monooxo bridging geometry, the $O_2^= \rightarrow M$ CT transitions appear[95] to shift up in energy by at least

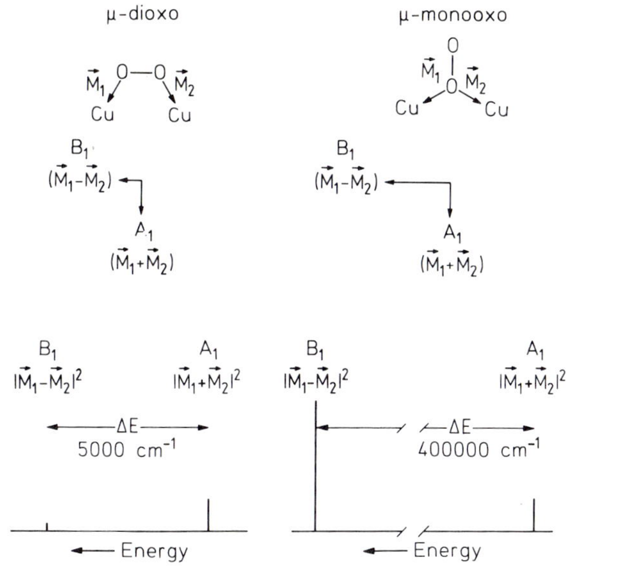


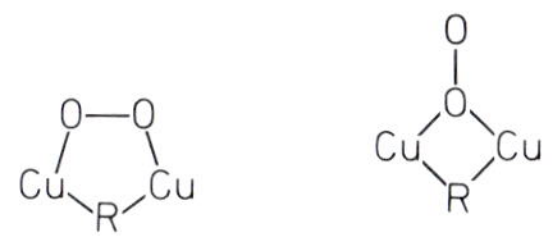


| | μ-dioxo | μ-monooxo | Oxyhemocyanin |
|---|---|---|---|
| $\pi_v^* \rightarrow d_{x^2-y^2}$ | | | |
| $I_{A_2}/I_{B_2}$ | 0.0 | 0.0 | 0.1 |
| $\Delta E(A_2-B_2)$ | 840 $cm^{-1}$ | 2400 $cm^{-1}$ | 3500 $cm^{-1}$ |
| $\pi_\sigma^* \rightarrow d_{x^2-y^2}$ | | | |
| $I_{A_1}/I_{B_1}$ | 2.6 | 0.24 | – |
| $\Delta E(A_1-B_1)$ | -5000 $cm^{-1}$ | -400000 $cm^{-1}$ | – |
| $\pi_\sigma^* - \pi_v^*$ | 12000 $cm^{-1}$ | 210000 $cm^{-1}$ | – |

**Fig. 35.** Diagrammatic explanation of the transition dipole vector coupling treatment for two possible peroxide bridging geometries, $\mu$-dioxo and $\mu$-monooxo, and comparison to oxyhemocyanin (see text for explanation)

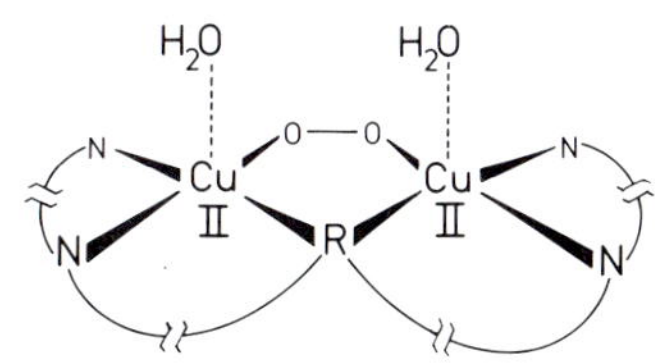


**Fig. 36.** Effective active site picture of oxyhemocyanin

4000 $cm^{-1}$. A detailed analysis of the binuclear cobalt(III) $\mu$-peroxo spectral analogues would provide valuable insight into the electronic structure of the oxy site. It would also test the generality of the transition dipole vector coupling model used in assigning the peroxide bridging geometry of oxy (*vide supra*) and lead to a better understanding of excited-state distortions of a coupled binuclear site.

A recent detailed spectroscopic study[96] has addressed an interesting question raised by resonance Raman data of the oxy coupled binuclear site (Fig. 6): no Cu–O metal ligand stretch is observed in resonance with the 350 nm $O_2^{=}\pi_\sigma^*(A_1) \rightarrow Cu(II)$ CT transition (the ~280 $cm^{-1}$ vibrations which are enhanced have been associated with Cu–N(His) stretches[97]). It is expected that excitation of an electron from a $\sigma$ bonding peroxide orbital into the antibonding Cu $d_{x^2-y^2}$ orbital would produce a significant distortion of the Cu–O bond. Resonance Raman studies of the met $N_3^-$ derivative[96] have helped to elucidate this question. Figure 37 shows the $N_3^- \rightarrow Cu(II)$ CT absorption and CD spectra of met $N_3^-$, as well as excitation profiles for the intra-azide stretch (2042 $cm^{-1}$) and the Cu–$N_3$ stretch (397 $cm^{-1}$). As described for oxy, at least three CT bands are required to fit this spectrum, indicating that azide must bridge the coppers. Further, a splitting of the intra-azide stretch with the mixed isotope, $^{14}N^{14}N^{15}N$, is also clearly observed in the spectrum of met azide hemocyanin. The resonance Raman profiles indicate that here, analogous to the oxy case, a copper-azide stretch is not in resonance with the $\pi_\sigma^* \rightarrow Cu(II)$ CT transition at 380 nm. Normal coordinate analysis of the met $N_3^-$ site shows that the absence of this feature is not associated with a significant mixing of this vibration into other normal modes, nor is it related to symmetry restrictions in the $C_{2v}$ binuclear site (the 397 $cm^{-1}$ vibration has $A_1$ symmetry as does the $N_3^-\pi_\sigma^* \rightarrow Cu(II)$ CT transition at 380 nm). Therefore, the lack of a Cu–O stretch in oxy appears to relate to the excited-state geometry of a $\sigma$ CT transition in a bridged binuclear com-

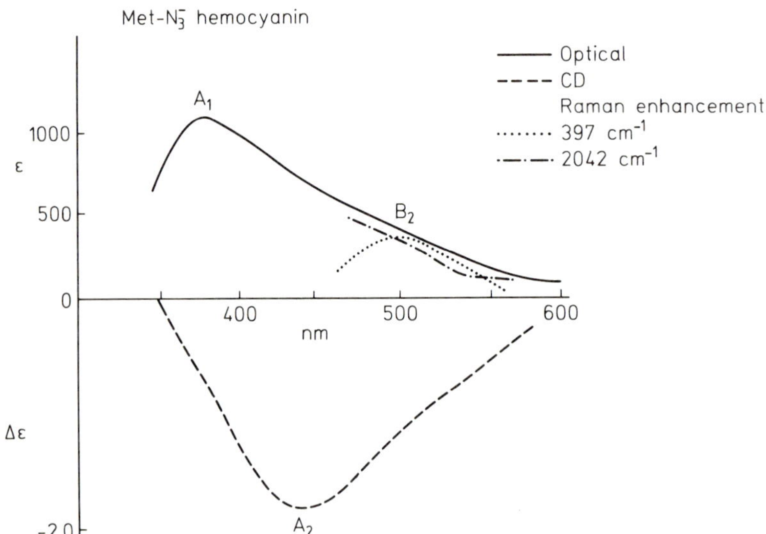


**Fig. 37.** $N_3^- \rightarrow Cu(II)$ charge transfer optical and CD spectra (room temperature) for met $N_3^-$ hemocyanin; excitation profiles[96] for the 2042 $cm^{-1}$ and 397 $cm^{-1}$ vibrations are indicated

plex. A normal coordinate treatment of the oxy site indicates that the $\sim 280\ \mathrm{cm}^{-1}$ $A_1$ vibration, which is enhanced by the $O_2^{=}\pi_\sigma^* \rightarrow$ Cu(II) CT transition, is in fact associated with the normal mode site distortion leading to a change in the Cu–Cu distance, indicated below:

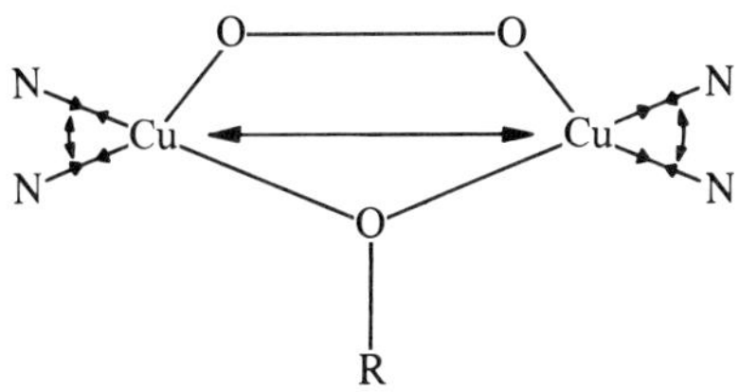


This point, however, must be investigated further on small-molecule spectral analogues; the $\mathrm{Co(III)}_2\mu\text{-}O_2^{=}$ complexes described earlier are logical candidates.

When the coupled binuclear copper site in the protein becomes structurally characterized by x-ray crystallography, single-crystal spectroscopy should provide a more detailed picture of its electronic structure and bonding, as has been achieved for the Blue Copper site described in the previous section.

# V. Active Sites in Multicopper Oxidases

The multicopper oxidases[32, 33] contain a combination of sites analogous to the three types discussed in the preceding sections (see Table 5): type 1 (Blue), type 2 (normal) and type 3 (coupled binuclear). These enzymes couple four one-electron oxidations of substrates to the four-electron reduction of dioxygen to water. The oxidases are the most complicated of copper-containing proteins, excluding cytochrome *c* oxidase, and the least understood. Nevertheless, significant progress is being made, particularly with respect to tree laccase, *Rhus vernicifera.* The approach described below parallels to some extent

**Table 5.** Distribution of Copper in Multicopper Oxidases

| Enzyme | | Number of sites per molecule | | |
|---|---|---|---|---|
| | | Type 1 (Blue) | Type 2 (normal) | Type 3 (coupled binuclear = met) |
| Laccase | Tree | 1 | 1 | 1 |
| | Fungal | 1 | 1 | 1 |
| Ceruloplasmin | | 2 | 1 | 1 or 2 |
| Ascorbate Oxidase | | 3 | 1 | 2 |

that which generated the current understanding of the coupled binuclear site (see Sect. IV). Derivatives have been prepared which reduce the number of coppers spectroscopically present. The properties of each site have then been separately defined and compared to the sites in Blue, normal, and coupled binuclear copper proteins; the possibilities of intersite interactions have also been probed.

Spectral studies of the native enzyme are difficult to interpret due to simultaneous contributions from all three types of copper. The EPR spectrum (Fig. 38) of native laccase[98] shows features attributable to the type 2 ($g_{\parallel} = 2.237$, $g_{\perp} = 2.053$, $A_{\parallel} = 206 \times 10^{-4}\ cm^{-1}$) and type 1 ($g_z = 2.300$, $g_y = 2.055$, $g_x = 2.030$, $A_z = 43 \times 10^{-4}\ cm^{-1}$) coppers. The type 3 site is EPR-non-detectable and diamagnetic ($-2J > 550\ cm^{-1}$)[56, 57]. This site should be compared to the met derivative of hemocyanin and tyrosinase only in the sense that it appears to contain two anti-ferromagnetically coupled copper(II)'s lacking an EPR signal. A characteristic "Blue" band of the type 1 copper is seen in the optical spectrum at 614 nm ($\varepsilon = 5700\ M^{-1}\ cm^{-1}$). The shoulder observed at 330 nm ($\varepsilon \approx 2800\ M^{-1}\ cm^{-1}$) was originally associated with the type 3 copper site, as it reduces with two electrons at

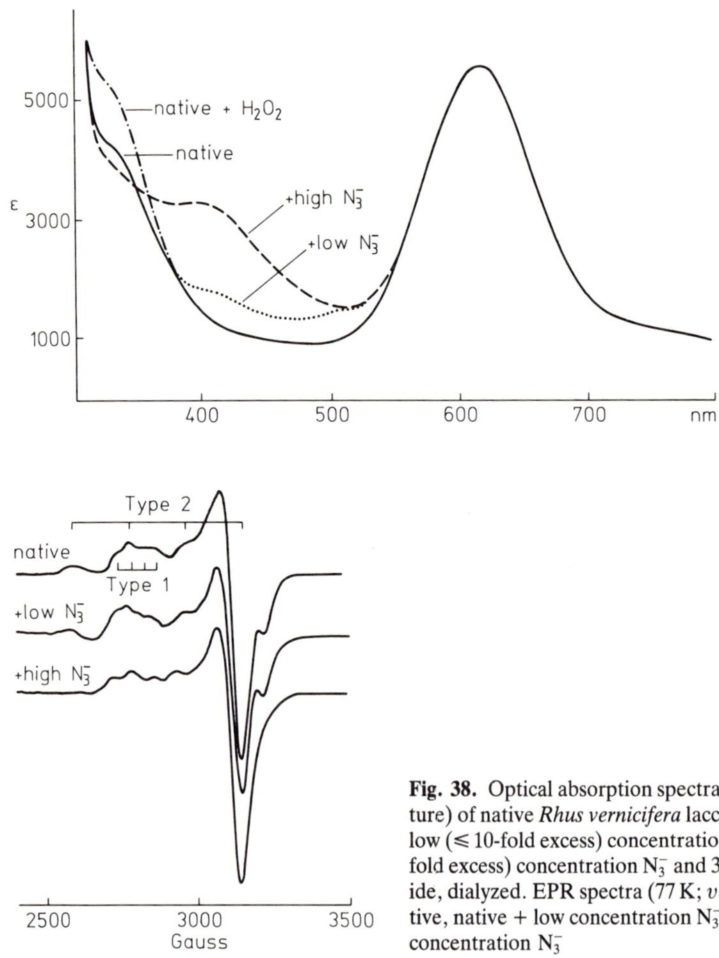


**Fig. 38.** Optical absorption spectra (room temperature) of native *Rhus vernicifera* laccase and native + low ($\leqslant$ 10-fold excess) concentration $N_3^-$, high ($\geqslant$ 100-fold excess) concentration $N_3^-$ and 30-fold excess peroxide, dialyzed. EPR spectra (77 K; $\upsilon = 9.1$ GHz) of native, native + low concentration $N_3^-$ and native + high concentration $N_3^-$

the same potential[99, 100]. Upon addition of low azide concentrations to the native enzyme at pH ≤ 6 (Fig. 38), $N_3^- \rightarrow Cu(II)$ CT transitions appear[101] at ~400 nm ($\varepsilon \approx 630\ M^{-1}\ cm^{-1}$) and 500 nm ($\varepsilon \approx 500\ M^{-1}\ cm^{-1}$). As the type 1 and 2 EPR are not significantly affected, these transitions are associated with the type 3 site. At high azide concentrations, an additional 400 nm ($\varepsilon \approx 1900\ M^{-1}\ cm^{-1}$) $N_3^- \rightarrow Cu(II)$ CT band grows in and the type 2 EPR signal changes. Stoichiometric addition of peroxide to native *Rhus* laccase at pH = 7 produces an increase in absorption[102] at ~330 nm ($\Delta\varepsilon \approx 800\ M^{-1}\ cm^{-1}$) with no change in the EPR spectrum. While this demonstrates that peroxide binds to native laccase with an extremely high affinity at pH ≥ 7, it is not clear whether these spectral changes are new $O_2^= \rightarrow Cu(II)$ CT transitions or perturbations of the existing 330 nm band (*vide infra*).

An important laccase derivative can be prepared[103] that has the type 2 copper removed (Fig. 39). The type 2 EPR features and the 330 nm absorption band are lost and $N_3^- \rightarrow Cu(II)$ CT transitions do not appear when azide is added to this type-2-depleted (T2D) derivative. Recent discussions have thus associated these spectral features (present only in native protein) with the type 2 copper. However, if T2D is treated[104] with excess peroxide, producing the treated type-2-depleted (TT2D) derivative, the 330 nm shoulder reappears, as do weak transitions at 410 nm and 745 nm. When azide is added to TT2D, an $N_3^- \rightarrow Cu(II)$ CT transition is observed at ~450 nm ($\varepsilon \approx 500\ M^{-1}\ cm^{-1}$). These results imply that the type 3 site in T2D is reduced and stable to aerobic oxidation but can be oxidized with excess peroxide. X-ray absorption data[104] (Fig. 40) corroborate this interpretation: only the T2D derivative shows a low-energy transition at 8984 eV (presently assigned as 1s → 4s) associated with Cu(I). Native and TT2D show identical x-ray absorption data which resemble those of Cu(II) imidazole complexes. This T2D derivative, which contains a reduced type 3 site, allows a clear

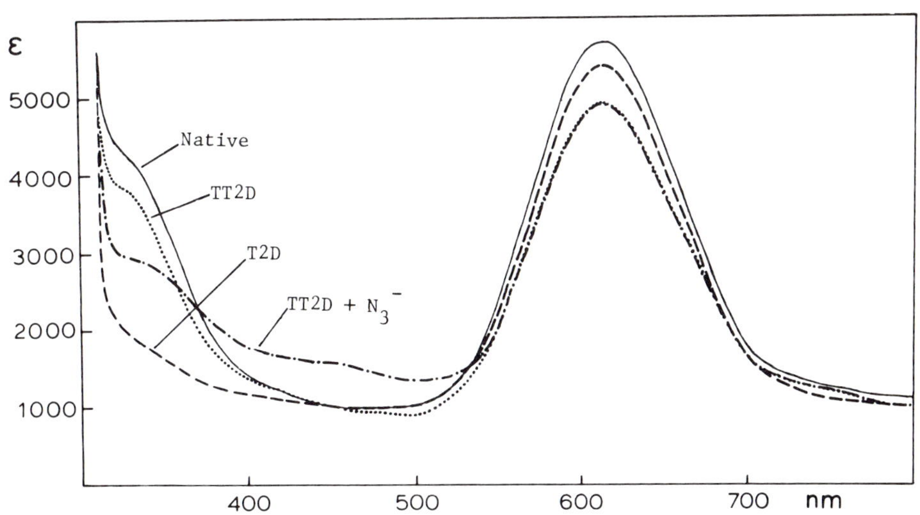


**Fig. 39.** Optical absorption spectra (room temperature) of native, type-2-depleted (T2D), T2D + 30-fold excess $H_2O_2$ (TT2D) and TT2D + 100-fold excess $N_3^-$ laccase (pH = 6.0; protein concentration ≈ 1.5 mM) (from Ref. 104)

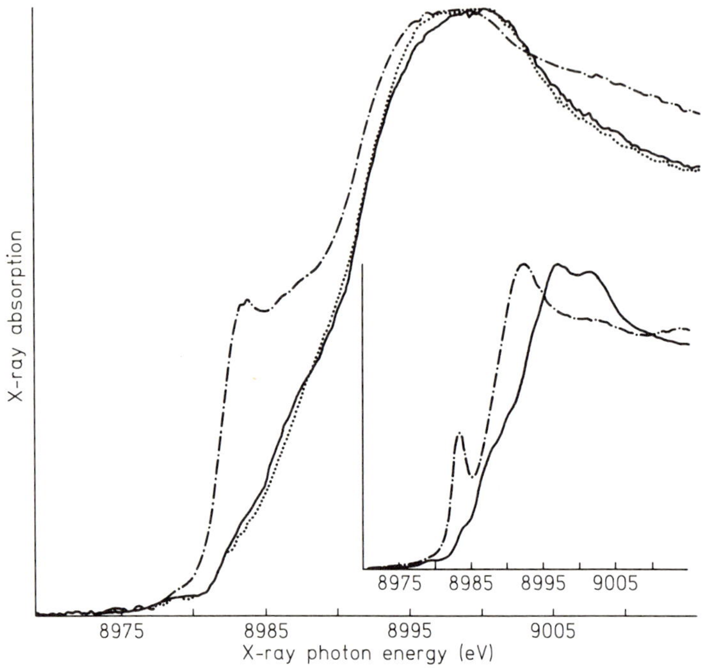


**Fig. 40.** X-ray absorption edges of native (——), T2D (–·–·–), and TT2D (·····) laccase (pH = 6.0; protein concentration = 5–6 mM). Insert: x-ray absorption edges of Cu(II) imidazole (——) and Cu(I) imidazole (–·–·–) complexes; all spectra taken at – 60 °C (from Ref. 104)

determination of the type 1 spectral features in laccase. These are $g_{\parallel} = 2.300$, $g_{\perp} = 2.054$, $A_{\parallel} = 37.8 \times 10^{-4}\,cm^{-1}$, $\lambda = 614$ nm ($\varepsilon = 5700\ M^{-1}\,cm^{-1}$), which are very similar to those of plastocyanin (see Sect. III) indicating comparable geometric and electronic structures.

Oxidation of the type 3 site in TT2D leads to the appearance of new spectral features (Fig. 39) with no increase in the EPR intensity. The type 3 optical features have thus been identified (330 nm, $\varepsilon = 2000\ M^{-1}\,cm^{-1}$; 420 nm, $\varepsilon = 175\ M^{-1}\,cm^{-1}$; 745 nm, $\varepsilon = 150\ M^{-1}\,cm^{-1}$)[104] and can now be correlated with those of the coupled binuclear copper site in the hemocyanins and tyrosinase. The 745 nm absorption can be assigned as d-d transitions of two tetragonal Cu(II)'s. For hemocyanin (which contains no strong sulfur ligation according to EXAFS data[82,83]), the 330 nm (broad shoulder with $\varepsilon \leq 1500\ M^{-1}\,cm^{-1}$ in met)[66] and 420 nm ($\varepsilon \geq 400\ M^{-1}\,cm^{-1}$ present in met and oxy) features have been assigned as histidine → Cu(II) and tyrosine → Cu(II) CT transitions, respectively. As tetragonal sulfur-to-copper(II) CT transitions can also contribute intensity in this region (Fig. 5) EXAFS data are required for the type 3 center in laccase before sulfur ligands can be excluded and firm analogous assignments made. Further, the appearance of an $N_3^- \rightarrow$ Cu(II) CT transition in TT2D (Fig. 39) indicates that azide does in fact bind to the oxidized type 3 copper site. The charge transfer absorption, CD, and associated resonance Raman spectra[105], however, differ significantly from that of met $N_3^-$ hemocyanin. This indicates a somewhat different small-molecule binding geometry at the type 3 site in TT2D. Further studies are required to elucidate this binding geometry; a stable half-met type 3 site in laccase has now been prepared[106, 107] and should be particularly useful for this purpose (see Note Added in Proof).

Upon oxidation of T2D, spectral changes[104] are also observed for the type 1 copper site (Fig. 41), indicating intersite interaction. The type 1 parallel hyperfine increases to $42.9 \times 10^{-4}$ cm$^{-1}$, intensity of the Blue band decreases ($\Delta\varepsilon_{614} \leqslant -300$ M$^{-1}$ cm$^{-1}$) and the resonance Raman spectrum of the Blue site shows a significant increase in intensity of a 382 cm$^{-1}$ vibration. These changes demonstrate that the geometry of the type 1 site is affected by oxidation of the type 3 copper in T2D laccase.

It is now possible to correlate this TT2D derivative with the native enzyme in order to determine the type 2 copper contributions. Type 2 EPR parameters can be accurately obtained in the overlapping perpendicular region ($g_\perp = 2.035$) and are consistent with tetragonal copper(II). Optically, tetragonal Cu(II) d-d transitions cannot be unambiguously distinguished in the complicated spectral region around 700 nm, but a difference between TT2D and native is observed in the 330 nm region (Fig. 39: $\Delta\varepsilon = 800$ M$^{-1}$ cm$^{-1}$); this could relate to histidine $\rightarrow$ tetragonal Cu(II) CT transitions associated with the type 2 site. However, the 330 nm region appears to be easily perturbed (as seen for native $N_3^-$ on TT2D and $O_2^=$ on native laccase) and must be further understood before these changes are meaningful.

Finally, both the type 2 and type 3 sites appear to be involved in binding exogenous small molecules to the active site in native laccase. Though peroxide oxidizes the reduced type 3 coppers in T2D, it is not bound with high affinity at this site. In the native enzyme, however, peroxide does bind strongly, producing the 330 nm spectral changes previously discussed. While not yet interpreted, this is clearly different from oxyhemocyanin and consistent with a different peroxide binding mode. Even though azide will bind to the TT2D protein, it binds to native laccase with a much higher affinity[101] and

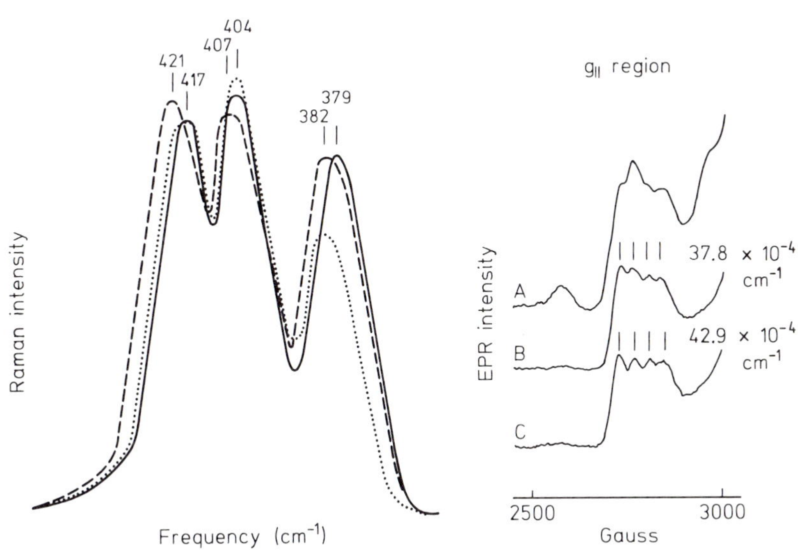


**Fig. 41.** Resonance Raman (room temperature; 100 mW irradiation at 568.2 nm) and EPR (77 K; $\upsilon = 9.1$ GHz) spectra of native (—, A), T2D (·····, B) and TT2D (---, C) laccase (pH = 6.0); $A_\parallel$ values are indicated (from Ref. 104)

quite different $N_3^- \rightarrow Cu(II)$ CT features (see Fig. 38 ··· and Fig. 39 –·–·–) when compared to both TT2D and met hemocyanin [native: $K \approx 10^4\ M^{-1}$ (410 nm and 500 nm); TT2D: $K \approx 10^2\ M^{-1}$ (450 nm); met hemocyanin: $K \approx 10^3\ M^{-1}$ (380 nm)]. At present the nature of the type 2 involvement in binding small molecules to the type 3 site is unclear. However, resonance Raman studies with mixed isotopes indicate[105] that any direct binding of azide to the type 2 copper must be relatively weak. A detailed investigation of these intersite interactions requires application of new spectral probes (in particular, variable temperature MCD and time-domain EPR) which will likely be emphasized in future multicopper enzyme studies.

## VI. Note Added in Proof

Recent EXAFS results[108] have excluded sulfur ligation at the T3 site in laccase, supporting assignment of the 330 nm region as dominantly deriving from a composite of histidine → Cu(II) charge transfer transitions. However, in contrast to the hemocyanins and tyrosinase, optical and EPR studies of half-met T2D laccase[109] have indicated that exogenous ligands do not appear to bridge the binuclear copper site. This is consistent with the significantly different azide spectral features of TT2D laccase compared to met $N_3^-$ hemocyanin. Similarly, we note[110] that the peroxylaccase optical features are relatively small, compared to the intense absorption bands observed upon binding $H_2O_2$ to met hemocyanin (Fig. 38 —— vs. –·–·, compared to Fig. 33 --- vs. ——), and indicative of a different $O_2^=$ binding mode in laccase. Hence, the apparent inability of exogenous ligands to bridge the T3 site suggests that peroxide binds as hydroperoxide to a single copper of the binuclear site in native laccase (Fig. 42); this likely relates to the irreversible binding of this two-electron reduced intermediate in the reduction of oxygen to water. However, comparative anion binding studies of native and TT2D laccase[111] indicate that the hydroperoxide does not bind strongly in this geometry without the presence of the T2 copper which apparently stabilizes the T3-hydroperoxide complex, perhaps through added positive charge in a low dielectric pocket.

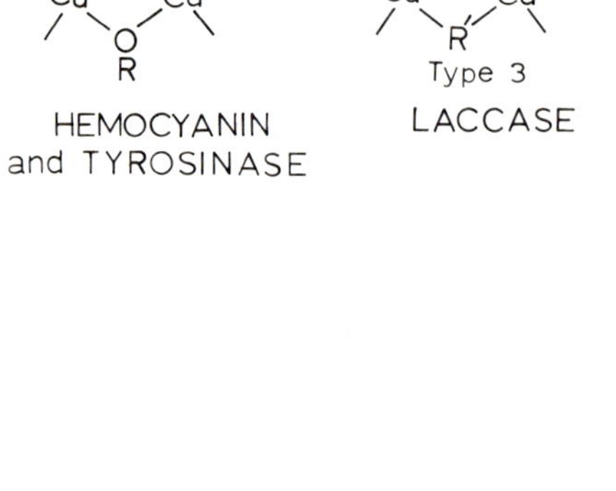


**Fig. 42.** Spectroscopically effective active site representations of the coupled binuclear copper site (*left*) and the type 3 site in *Rhus* laccase (*right*) where OR and R′ represent endogenous protein bridges in the respective sites

*Acknowledgements.* We are grateful to the National Institute of Arthritis, Diabetes, and Digestive and Kidney Diseases of the U.S. Public Health Department (Grant # AM 31450) and the National Science Foundation (Grant # PCM 78-13714) for supporting different phases of our research. EIS would like to thank his students Nancy Eickman, Robert Gay, Richard Himmelwright, Cynthia LuBien, Darlene Spira, Thomas Thamann and Marjorie Winkler, and his collaborators Hans Freeman, Harry Gray, Keith Hodgson, Konrad Lerch, William Mims, Harvey Schugar, Thomas Spiro and Philip Stephens, who have contributed to various portions of the work described in this review.

# VII. References

1. Gray, H. B., Solomon, E. I.: Electronic structure of blue copper centers in proteins, in: Copper Proteins (ed.) Spiro, T. G., chapter 1, New York, Wiley-Interscience 1981
2. Solomon, E. I.: The binuclear copper active site: hemocyanin, tyrosinase and type 3 copper oxidases, *ibid.*, chapter 2
3. (a) Ballhausen, C. J.: Introduction to Ligand Field Theory, New York, McGraw-Hill 1962
   (b) Mulliken, R. S., Rieke, C. A.: Repts. Progr. Phys. *8,* 231 (1941)
4. (a) Abragam, A., Bleaney, B.: Electron Paramagnetic Resonance of Transition Ions, Oxford, Clarendon 1970
   (b) Maki, A., McGarvey, B.: J. Chem. Phys. *29,* 31 (1958)
5. (a) Peisach, J., Blumberg, W. E.: Arch. Biochem. Biophys. *165,* 691 (1974)
   (b) Walker, F. A., Sigel, H., McCormick, D. B.: Inorg. Chem. *11,* 2756 (1972)
   (c) Falk, K.-E. et al.: Inorg. Chem. *3,* 556 (1970)
   (d) Lewis, W. B., Alei, M., Morgan, L. O.: J. Chem. Phys. *45,* 4003 (1966)
   (e) Malmström, B. G., Vänngård, T.: J. Mol. Biol. *2,* 118 (1960)
   (f) Davis, P. H., White, L. K., Belford, R. L.: Inorg. Chem. *14,* 1753 (1975). See discussion in caption of Fig. 5.
   (g) Chow, C., Chang, K., Willett, R. D.: J. Chem. Phys. *59,* 2629 (1973)
6. Froncisz, W. et al.: J. Biol. Chem. *254,* 7482 (1979)
7. Feher, G. et al.: Ann. N.Y. Acad. Sci. *222,* 86 (1973)
8. Mims, W. B., Peisach, J.: Pulsed EPR studies of metalloproteins, in: Biological Applications of Magnetic Resonance (ed.) Schulman, R. G., p. 221, New York, Academic, 1979
9. Mims, W. B.: The Linear Electric Field Effect in Paramagnetic Resonance, Oxford, Clarendon 1976
10. Peisach, J., Mims, W. B.: Chem. Phys. Lett. *37,* 307 (1976)
11. (a) Hathaway, B. J., Billing, D. E.: Coord. Chem. Rev. *5,* 143 (1970)
   (b) Holmes, O. G., McClure, D. S.: J. Chem. Phys. *26,* 1686 (1957)
   (c) Hathaway, B. J., Stephens, F.: J. Chem. Soc. A, 884 (1970)
   (d) McFadden, D. L. et al.: J. Chem. Soc. Dalton Trans. 47 (1976)
   (e) McFaden, D. L. et al.: J. Chem. Soc. Dalton Trans. 263 (1975)
   (f) McLellan, A. W., Melson, G. A.: J. Chem. Soc. A, 137 (1967)
   (g) Ferris, N. S. et al.: J. Am. Chem. Soc. *100,* 5939 (1978)
   (h) Glick, M. D. et al.: Inorg. Chem. *15,* 1190 (1976)
   (i) Miskowski, V. M. et al.: J. Am. Chem. Soc. *98,* 8344 (1976)
   (j) Ou, C. C. et al.: Inorg. Chem. *15,* 3157 (1976)
   (k) Hitchman, M. A., Cassidy, P. J.: Inorg. Chem. *18,* 1745 (1979)
   (l) Oelkrug, D.: Struct. Bonding *9,* 1 (1971)
   (m) Oelkrug, D.: Z. Physik. Chem. N.F. *56,* 325 (1967)
   (n) Kato, H.: Mol. Phys. *24,* 81 (1972)
   (o) Yasui, T., Hidaka, J., Shimura, Y.: J. Am. Chem. Soc. *87,* 2762 (1965)
12. (a) Amundsen, A. R., Whelan, J., Bosnich, B.: J. Am. Chem. Soc. *99,* 6730 (1977)
   (b) Elliot, H., Hathaway, B. J.: Inorg. Chem. *5,* 885 (1966)

(c) Kennedy, B. P., Lever, A. B. P.: J. Am. Chem. Soc. *95,* 6907 (1973)
(d) Bernarducci, E. et al.: J. Am. Chem. Soc. *103*, 1686 (1981)
(e) Fawcett, T. G. et al.: J. Am. Chem. Soc. *102,* 2598 (1980)
(f) Downes, J. M., Whelan, J., Bosnich, B.: Inorg. Chem. *20,* 1081 (1981)
(g) Hughey, J. L. IV et al.: J. Am. Chem. Soc. *101,* 2617 (1979)
(h) Agnus, Y., Louis, R., Weiss, R.: J. Am. Chem. Soc. *101,* 3381 (1979)

13. Ballhausen, C. J.: The measurement and interpretation of transition ion crystal spectra, in: Spectroscopy in Inorganic Chemistry, Vol. 1 (eds.) Rao, C. N. R., Ferraro, J. R., chapter 1, New York, Academic 1970
14. (a) Gillard, R. D.: Optical rotatory dispersion and circular dichroism, in: Physical Methods in Advanced Inorganic Chemistry (eds.) Hill, H. A. O., Day, P., chapter 5, New York, Wiley-Interscience 1968
    (b) Richardson, F. S.: Chem. Rev. *79,* 17 (1979)
15. Kuhn, W.: Ann. Rev. Phys. Chem. *9,* 417 (1958)
16. (a) Stephens, P. J.: Adv. Chem. Phys. *35,* 197 (1976)
    (b) Stephens, P. J.: Ann. Rev. Phys. Chem. *25,* 201 (1974)
17. Eickman, N. C. et al.: J. Am. Chem. Soc. *100,* 6529 (1978)
18. Freedman, T. B., Loehr, J. S., Loehr, T. M.: J. Am. Chem. Soc. *98,* 2809 (1976)
19. Larrabee, J. A. et al.: J. Am. Chem. Soc. *99,* 1979 (1977)
20. Valentine, J. S., Pantoliano, M. W.: Protein-metal ion interactions in cuprozinc protein (superoxide dismutase), in: Copper Proteins (ed.) Spiro, T. G., chapter 8, New York, Wiley-Interscience 1981
21. Ettinger, M. J., Kosman, D. J.: Chemical and catalytic properties of galactose oxidase, *ibid.,* chapter 6
22. Villafranca, J. J.: Dopamine-$\beta$-hydroxylase, *ibid.,* chapter 7
23. Hamilton, G. A.: Oxidases with monocopper reactive sites, *ibid.,* chapter 5
24. Richardson, J. S. et al.: Proc. Nat. Acad. Sci. USA *72,* 1349 (1975)
25. Weser, U. et al.: Biochim. Biophys. Acta *243,* 203 (1971)
26. Rotilio, G. et al.: Biochemistry *11,* 2187 (1972)
27. Lieberman, R. A., Sands, R. H., Fee, J. A.: J. Biol. Chem. *257,* 336 (1982)
28. Tainer, J. A. et al.: in 250D: Cu,Zn-Superoxide Dismutase Complete Atomic Coordinates (eds.) Richardson, D. C. and Richardson, J. S., Brookhaven Protein Structure Data Bank 1980
29. Rotilio, G., Calabrese, L., Coleman, J. E.: J. Biol. Chem. *248,* 3855 (1973)
30. Weser, U.: Struct. Bonding (Berlin) *17,* 1 (1973)
31. (a) Fee, J. E., Gaber, B. P.: J. Biol. Chem. *247,* 60 (1972)
    (b) Rotilio, G., Rigo, A., Calabrese, L.: Recent developments on the active-site structure and mechanism of bovine copper- and zinc-containing superoxide dismutase, in: Frontiers in Physiochemical Biology (ed.) Pullman, B., chapter 15, New York, Academic 1978
    (c) Fee, J. A., Peisach, J., Mims, W. B.: J. Biol. Chem. *256,* 1910 (1981)
32. Malkin, R., Malmström, B. G.: The state and function of copper in biological systems, in: Advances in Enzymology and Related Areas of Molecular Biology, 33 (ed.) Nord, F. F., p. 177, New York, Interscience 1970
33. Fee, J. A.: Struct. Bonding (Berlin) *23,* 1 (1975)
34. Solomon, E. I., Hare, J. W., Gray, H. B.: Proc. Nat. Acad. Sci. USA *73,* 1389 (1976)
35. Solomon, E. I. et al.: J. Am. Chem. Soc., *102,* 168 (1980)
36. Osborne, G. H., Cheng, J. C., Stephens, P. J.: Rev. Sci. Instrum. *44,* 10 (1973)
37. (a) Bates, C. A. et al.: Proc. Phys. Soc. *79,* 73 (1962)
    (b) Parker, I. H.: J. Phys. C.: Solid St. Phys. *4,* 2967 (1971)
    (c) Krishnan, V. G.: J. Chem. Phys. *68,* 660 (1978)
    (d) Bates, C. A.: Proc. Phys. Soc. *83,* 465 (1964)
    (e) Dietz, R. E. et al.: Phys. Rev. *132,* 1559 (1963)
38. Roberts, J. E. et al.: J. Am. Chem. Soc. *102,* 825 (1980)
39. Katoh, S., Takamiya, S.: J. Biochem. (Tokyo) *55,* 378 (1964)
40. Solomon, E. I. et al.: J. Am. Chem. Soc. *97,* 3878 (1975)

41. (a) McMillan, D. R., Holwerda, R. A., Gray, H. B.: Proc. Nat. Acad. Sci. USA *71,* 1339 (1974)
(b) McMillan, D. R., Rosenberg, R. C., Gray, H. B.: Proc. Nat. Acad. Sci. USA *71,* 4760 (1974)
(c) Solomon, E. I. et al.: J. Am. Chem. Soc. *98*, 8046 (1976)
42. Colman, P. J. et al.: Nature (London) *272,* 319 (1978)
43. Markley, J. L.: Biochemistry *14,* 4428 (1975)
44. Penfield, K. W. et al.: J. Am. Chem. Soc. *103,* 4382 (1981)
45. Desjardins, S. R. et al.: to be published
46. (a) Tennent, D. L., McMillan, D. R.: J. Am. Chem. Soc. *101*, 2307 (1979)
(b) McMillan, D. R., Morris, M. C.: Proc. Nat. Acad. Sci. USA *78,* 6567 (1981)
47. Peisach, J., Levine, W. G., Blumberg, W. E.: J. Biol. Chem. *242,* 2847 (1967)
48. Companion, A. L., Komarynsky, M. A.: J. Chem. Educ. *41,* 257 (1964)
49. Peisach, J., Mims, W. B.: Eur. J. Biochem. *84*, 207 (1978)
50. Bannister, J. V. (ed.): Structure and Function of Haemocyanin, Berlin, Springer-Verlag 1977
51. Ghiretti, F.: Arch. Biochem. Biophys. *63,* 165 (1956)
52. Himmelwright, R. S. et al.: J. Am. Chem. Soc. *102,* 5378 (1980)
53. Himmelwright, R. S. et al.: J. Am. Chem. Soc. *102,* 7339 (1980)
54. van Schaick, E. J. M. et al.: The crystal structure of hemocyanin hexamer from *Panulirus interruptus* at 0.5 nm resolution, in: Invertebrate Oxygen-Binding Proteins (eds.) Lamy, J. and Lamy, J., p. 353, New York, Marcel Dekker 1981
55. Magnus, K. A., Love, W. E.: Crystals of a 70,000 dalton subunit of *Limulus polyphemus* hemocyanin, *ibid.*, p. 363
56. Solomon, E. I. et al.: J. Am. Chem. Soc. *98,* 1029 (1976)
57. Dooley, D. et al.: Proc. Nat. Acad. Sci. USA *75,* 3019 (1978)
58. Schugar, H. et al.: J. Am. Chem. Soc. *97,* 6442 (1975)
59. Hansen, A. E., Ballhausen, C. J.: Trans. Faraday Soc. *61,* 631 (1965)
60. Smith, T. D., Pilbrow, J. R.: Coord. Chem. Rev. *13,* 173 ( 1974)
61. Himmelwright, R. S., Eickman, N. C., Solomon, E. I.: Biochem. Biophys. Res. Comm. *81,* 237 (1978)
62. Himmelwright, R. S., Eickman, N. C., Solomon, E. I.: Biochem. Biophys. Res. Comm. *81,* 243 (1978)
63. Himmelwright, R. S., Eickman, N. C., Solomon, E. I.: Biochem. Biophys. Res. Comm. *84,* 300 (1978)
64. Himmelwright, R. S., Eickman, N. C., Solomon, E. I.: J. Am. Chem. Soc. *101,* 1576 (1979)
65. Himmelwright, R. S., Eickman, N. C., Solomon, E. I.: Biochem. Biophys. Res. Comm. *86,* 628 (1979)
66. Eickman, N. C., Himmelwright, R. S., Solomon, E. I.: Proc. Nat. Acad. Sci. USA *76,* 2094 (1979)
67. Robin, M. B., Day, P.: Adv. Inorg. Chem. and Radiochem. *10,* 247 (1967)
68. Wilcox, D. E., Mims, W. B., Solomon, E. I.: to be published
69. Fenton, D. E., Schroeder, R. R., Lintvedt, R. L.: J. Am. Chem. Soc. *100,* 1931 (1978)
70. Gagné, R. R. et al.: J. Am. Chem. Soc. *101*, 4571 (1979)
71. Gatteschi, D., Mealli, C., Sacconi, L.: Inorg. Chem. *15,* 2774 (1976)
72. Gagné, R. R., Henling, L. M., Kistenmacher, T. J.: Inorg. Chem. *19,* 1226 (1980)
73. van der Deen, H., Hoving, H.: Biochemistry *16,* 3519 (1977)
74. (a) Schoot Uiterkamp, A. J. M.: FEBS Lett, *20,* 93 (1972)
(b) Schoot Uiterkamp, A. J. M., Mason, H. S.: Proc. Nat. Acad. Sci. USA *70,* 993 (1973)
75. Schoot Uiterkamp, A. J. M. et al.: Biochim. Biophys. Acta *372,* 407 (1974)
76. These simulation programs, based on a perturbation approach, have been used to estimate the metal-metal distance for a number of dipolar-coupled ($|2J| < 30\ cm^{-1}$) cupric complexes (Boyd, P. D. W. et al.: J. Chem. Soc. Dalton, 1549 (1973)) and show good agreement when compared with x-ray crystallographic results
77. Witters, R., Lontie, R.: FEBS Lett. *60,* 400 (1965)
78. Felsenfeld, G., Printz, M. P.: J. Am. Chem. Soc. *81,* 6259 (1959)

79. Witters, R., DeLey, M., Lontie, R.: The reaction of *Helix pomatia* methemocyanin with azide and fluoride, in: Structure and Function of Haemocyanin (ed.) Bannister, J. V., p. 239, Berlin, Springer-Verlag 1977
80. Wilcox, D. E., Long, J. R., Solomon, E. I.: to be published
81. Powers, L. et al.: to be published
82. Brown, J. M. et al.: J. Am. Chem. Soc. *102,* 4210 (1980)
83. Co, M. S. et al.: J. Am. Chem. Soc. *103,* 984 (1981)
84. (a) Hatfield, W. E.: Superexchange interactions in copper(II) complexes, in: Extended Interactions Between Metal Ions in Transition Metal Complexes (ed.) Interrante, L. V., chapter 10, Washington, D. C., ACS, 1974
(b) Coughlin, P. K., Lippard, S. J.: J. Am. Chem. Soc. *103,* 3228 (1981)
(c) Arcus, C., Fivizzani, K. P., Pankovic, S. F.: J. Inorg. Nucl. Chem. *39,* 285 (1977)
(d) Ishimura, Y.: Bull. Chem. Soc. Japan *46,* 3728 (1973)
(e) Loeb, S. J., Martin, J. W. L., Willis, C. J.: Inorg. Chem. *18,* 3160 (1979)
(f) Timmons, J. H. et al.: Inorg. Chem. *20,* 181 (1981)
(g) Curtis, N. F. et al.: J. Chem. Soc. Dalton, 1051 (1977)
(h) Dickson, I. E., Robson, R.: Inorg. Chem. *13,* 1301 (1974)
(i) Countryman, R. M., Robinson, W. T., Sinn, E.: Inorg. Chem. *13,* 2013 (1974)
(j) Harris, C. M., Sinn, E.: J. Inorg. Nucl. Chem. *30,* 2723 (1968)
(k) Gruber, S. J., Harris, C. M., Sinn, E.: J. Inorg. Nucl. Chem. *30,* 1805 (1968)
85. Goodenough, J. B.: Magnetism and the Chemical Bond, New York, Interscience 1963
86. (a) Wong, R. Y., Palmer, K. J., Tomimatsu, Y.: Acta. Cryst. *B32,* 567 (1976)
(b) Tomimatsu, Y., Kint, S., Scherer, J. R.: Biochemistry *15,* 4918 (1976)
(c) Harrod, J. F.: Canad. J. Chem. *47,* 637 (1969)
(d) Bullock, J. I., Hobson, R. J., Povey, D. C.: J. Chem. Soc. Dalton 2037 (1974)
(e) Ladd, M. F. C., Povey, D. C.: J. Cryst. Mol. Str. *6,* 205 (1976)
(f) Hobson, R. J., Ladd, M. F. C., Povey, D. C.: J. Cryst. Mol. Str *3,* 377 (1973)
(g) Frost, A. E. et al.: J. Am. Chem. Soc. *80,* 530 (1958)
(h) Boggess, R. K., Martin, R. B.: J. Am. Chem. Soc. *97,* 3076 (1975)
(i) Ainscough, E. W., Brodie, A. M., Plowman, J. E.: Inorg. Chim. Acta. *33,* 149 (1979)
(j) Letter, J. E. Jr., Bauman, J. E. Jr.: J. Am. Chem. Soc. *92,* 443 (1970)
(k) Letter, J. E. Jr., Bauman, J. E. Jr.: J. Am. Chem. Soc. *92,* 437 (1970)
(l) Ainscough, E. W. et al.: Biochemistry *19,* 4072 (1980)
87. The $B_{\bar{1}}$ component appears to have been observed in the UV-CD difference spectrum of oxy versus met at ~260 nm: K. Lerch, private communication
88. Thewalt, U., Struckmeier, G.: Z. anorg. allg. Chem. *419,* 163 (1976)
89. Thewalt, U.: Z. anorg. allg. Chem. *393,* 1 (1972)
90. Zehnder, M., Thewalt, U., Fallab, S.: Helv. Chim. Acta *59,* 2290 (1976)
91. Thewalt, U., Marsh, R.: J. Am. Chem. Soc. *89,* 6364 (1967)
92. Lever, A. B. P., Gray, H. B.: Acc. Chem. Res. *11,* 348 (1978)
93. Sasaki, Y., Fujita, J., Saito, K.: Bull. Chem. Soc. Japan *44,* 3373 (1971)
94. Fritch, J. R., Christoph, G. G., Schaefer, W. P.: Inorg. Chem. *12,* 2170 (1973)
95. Mori, M., Weil, J. A.: J. Am. Chem. Soc. *89,* 3732 (1967)
96. Thamann, T. J., Solomon, E. I.: to be published
97. Larrabee, J. A., Spiro, T. G.: J. Am. Chem. Soc. *102,* 4217 (1980)
98. Reinhammar, B., Malmström, B. G.: "Blue" copper-containing oxidases, in Copper Proteins (ed.) Spiro, T. G., chapter 3, New York, Wiley-Interscience 1981
99. Reinhammar, B. R. M., Vänngård, T. I.: Eur. J. Biochem. *18,* 463 (1971)
100. Reinhammar, B.: Biochim. Biophys. Acta *275,* 245 (1972)
101. Morpurgo, L. et al.: Biochim. Biophys. Acta *336,* 324 (1974)
102. Farver, O., Goldberg, J., Pecht, I.: Eur. J. Biochem. *104,* 71 (1980)
103. Graziani, M. T. et al.: FEBS Lett. *70,* 87 (1976)
104. LuBien, C. D. et al.: J. Am. Chem. Soc. *103,* 7014 (1981)
105. Thamann, T. J., LuBien, C. D., Solomon, E. I.: to be published
106. Spira, D. J., Solomon, E. I.: to be published
107. A half-met laccase kinetic intermediate has also been reported: Reinhammar, B.: J. Inorg. Biochem. *15,* 27 (1981)

108. Spira, D. J., Co, M. S., Hodgson, K. O., Solomon, E. I.: to be published
109. Spira, D. J., Winkler, M. E., Solomon, E. I.: Biochem. Biophys. Res. Comm. *107*, 721 (1982)
110. Winkler, M. E., Spira, D. J., LuBien, C. D., Thamann, T. J., Solomon, E. I.: *ibid.*, 727 (1982)
111. LuBien, C. D., Solomon, E. I., Spira, D. J., Thamann, T. J., Winkler, M. E.: manuscript in preparation

# Fe-S and Mo-Fe-S Clusters as Models for the Active Site of Nitrogenase

**Bruce A. Averill**

Department of Chemistry, Michigan State University, East Lansing, Michigan 48824, USA

Studies on the enzyme nitrogenase have shown that it contains two novel metal-sulfur clusters: the iron-molybdenum cofactor (FeMo-co), which contains one Mo, 6–7 Fe, and ~4 S atoms in a spin-coupled paramagnetic (S = 3/2) cluster; and the P clusters, variants of normal but highly reduced 4 Fe-4 S clusters in which three Fe atoms are somehow distinguished from the fourth and which, when oxidized, exhibit an usually high spin (S = 5/2 or 7/2) ground state. No structures consistent with all available spectroscopic data have been proposed for either unit. Synthetic efforts aimed at developing models for FeMo-co have utilized tetrathiomolybdate as starting material, and have resulted in two well-characterized classes of Mo-Fe-S cluster: those containing the $MoFe_3S_4$ "cubane" core and those containing a "linear" $MoS_2Fe$ unit. The former apparently contain Mo in a relatively reduced valence state (*ca.* + 3) and exhibit Mo-S and Mo-Fe distances in good agreement with those found for the Mo site of nitrogenase; the latter contain Mo in a higher oxidation state (*ca.* + 6) and have significantly shorter Mo-S distances. Synthetic work related to the P clusters has centered on the properties of the $[Fe_4S_4(SR)_4]^{3-}$ ions, for which two different structures with different magnetic properties are found, and on the properties of 4 Fe-4 S clusters with non-thiolate ligands.



Structure and Bonding 53

## Abbreviations

| | |
|---|---|
| ATP | adenosine triphosphate |
| DMF | N,N-dimethylformamide |
| DMSO | dimethylsulfoxide |
| EDTA | ethylenediamine-N,N,N',N'-tetra-acetate |
| EPR | electron paramagnetic resonance |
| EXAFS | extended X-ray absorption fine structure |
| FeMo-co | iron-molybdenum cofactor |
| MCD | magnetic circular dichroism |
| MeCN | acetonitrile |
| NMF | N-methylformamide |
| SCE | standard calomel electrode |
| SHE | standard hydrogen electrode |

# I. Introduction

Of the relatively well-characterized soluble metalloenzymes, the enzyme nitrogenase appears to have a special fascination for the inorganic chemist, combining as it does the ability to reduce $N_2$ at ambient temperature and pressure with the presence of novel metal-sulfur clusters. It is now abundantly clear that complete understanding of the details of the nitrogenase system will require that results from several areas of inorganic chemistry, including dinitrogen binding and activation[1], molybdenum chemistry[2], and iron-sulfur cluster chemistry[3–5], be brought to bear on the problem, in addition to more traditional biochemical and enzymological approaches.

Each of the above topics comprises a separate and mature subfield of inorganic chemistry, as evidenced by the authoritative review articles referenced above. With regard to biological aspects of the problem, both molybdenum[6–8] and iron-sulfur proteins[5, 9–12] have been the subject of extensive reviews and entire monographs, while a number of recent reviews[13–16] and monographs[17–19] and a series of proceedings of international symposia[20–23] have been devoted to the specific problem of biological nitrogen fixation and the enzyme system responsible for the first step, the reduction of dinitrogen to ammonia.

This review will focus on recent developments in the chemistry of iron-sulfur and molybdenum-iron-sulfur clusters that are aimed at producing models for the active site of nitrogenase. In view of the paucity of firm structural data on the enzyme system, it is clear that the compounds prepared to date are best classified as "speculative models" rather than "corroborative models"[24, 25]. That is, we have only a rough idea of the composition and some of the physical properties of the system for which we wish to construct models. Because of this and the difficulty involved in designing a rational synthesis for a cluster of specified stoichiometry and structure, it is somewhat premature to expect to prepare a synthetic cluster that is congruent in all of its essential physical properties with any of the metal centers of the enzyme. Instead, inorganic chemists are in the position of very recently having had it called to their attention by biochemists and biophysicists that nitrogenase utilizes metal clusters of unusual composition and with novel physical properties. We are thus just beginning to explore what has already proven to be a broad and fruitful area of research. This review is intended to stimulate new ideas and to encourage others to attack this important problem, In addition to contributing to our understanding of biological nitrogen fixation, research in this area has an important long range practical goal: the development of synthetic Mo-Fe-S catalysts that incorporate some of the essential features of the enzymatic system for reduction of dinitrogen to ammonia.

# II. Nitrogenase

## A. *The Nitrogenase System*

The enzyme nitrogenase is found in both symbiotic and free-living prokaryotic microorganisms that grow either aerobically or anaerobically. Reduction of nitrogen by the enzyme is a reductive dephosphorylation; at least four ATP molecules are hydrolyzed for every two electrons transferred to substrate[26]. In all cases investigated to date, the active enzyme consists of two oxygen-sensitive components: the Fe protein (mol. wt. ~60,000; 4 Fe and 4 $S^=$/mol) and the Mo-Fe protein (mol. wt. ~220,000; ~32 Fe, 32 $S^=$ and 2 Mo/mol)[14]. In certain organisms (such as *Rhodospirillum rubrum*), the Fe protein as isolated is inactive and must be activated by a separate activating factor[27] that apparently removes a covalently bound adenine-like molecule[28].

The available evidence[26] seems to indicate that electrons are transferred from external reductant (reduced ferredoxin or dithionite) to an MgATP complex of the Fe protein to the MoFe protein to substrate; in other words, it is the MoFe component that actually reduces $N_2$. In fact, based on these results and on more detailed kinetics studies[29], Burris has proposed that MoFe protein should be termed dinitrogenase, and the Fe protein dinitrogenase reductase. Although this terminology is by no means universally accepted, it supports the idea that it is at the metal centers of the MoFe protein that substrates (including $N_2$) are reduced.

A schematic, highly simplified view of electron transfer through the components and metal centers of nitrogenase is shown in Fig. 1. Although almost certainly not correct in all details, Fig. 1 does illustrate certain essential points: electron transfer is from the $(MgATP)_2$ complex of the Fe protein to the MoFe protein to substrate; the functional subunit of the MoFe protein is probably the half molecule containing two polypeptide chains, two P clusters, and one FeMo-cofactor (M center); substrates such as $N_2$ are probably reduced at the FeMo-cofactor, with the P-clusters acting primarily as electron conduits or reservoirs. Each of the component proteins is discussed below, with special emphasis on available data regarding the structures and roles of the metal-sulfur chromophores.

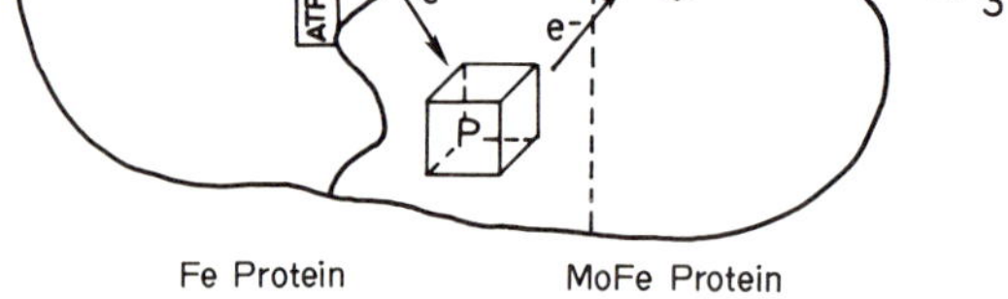


**Fig. 1.** Hypothetical diagram of the nitrogenase system. Electron transfer is from reduced ferredoxin or dithionite (not shown) to the Fe protein · $(MgATP)_2$ complex to the MoFe protein, with concomitant hydrolysis of the MgATP. Within the MoFe protein, electron transfer is shown as proceeding sequentially via either P cluster to the M center (FeMo-co), but neither this configuration nor one in which the P clusters are arranged in a linear fashion between the site of electron transfer and the M center is completely satisfactory. Nitrogen independent hydrogen evolution might occur at the very highly reduced P clusters, by analogy to the hydrogenase system[196–199], but evidence for this is not compelling. Nitrogen reduction is shown as occurring at the M center, but again there is no direct evidence for this

## B. The Iron Protein

In all systems examined, the iron protein of nitrogenase appears to consist of two identical subunits of molecular weight 27,000–34,000 and contains 3–4 atoms each of iron and of acid-labile sulfide per molecule[14)]. These data, combined with the fact that the Fe protein binds two MgATP per molecule[30–32)] and an early controversy as to whether one or two electrons were transported per molecule, made an unambiguous determination of the nature of the Fe-S chromophore(s) essential. Cluster displacement (extrusion) experiments in two laboratories[14, 33)] have shown that >95% of the iron in the iron protein is present as a 4 Fe–4 S cluster of the type present in the bacterial ferredoxins. This cluster presumably shuttles between the net 2+ and 1+ core oxidation states, corresponding to oxidized and reduced bacterial ferredoxins, respectively. This is consistent with the rhombic g = 1.94 EPR signal observed upon reduction[14)]. In conjunction with the observation[29, 34)] that MgATP is hydrolyzed only during transfer of an electron from the Fe protein to the MoFe protein, the above results suggest that the role of the Fe protein is to function as a transducer, utilizing the energy derived from hydrolysis of ATP to destabilize the reduced (1+) 4 Fe–4 S cluster, resulting in delivery of a very low potential ($\ll -400$ mV) electron to the MoFe protein. The chemical means by which this is accomplished are not clear. It is known that binding of MgATP to the Fe protein results in a change of the EPR spectrum from rhombic to axial[32, 35)], as well as a negative shift in reduction potential (from $-294 \pm 20$ mV to $-402 \pm 22$ mV[35)]). The perturbation of the EPR spectrum upon binding MgATP has generally been attributed to a change in protein conformation[14)]. Although the shift in reduction potential is in the expected direction, the fact that MgADP also binds to the Fe protein and induces a comparable shift in reduction potential suggests that this cannot provide the driving force for nitrogen fixation. An alternative involving phosphorylation of one (or two) of the terminal cysteinyl mercaptide ligands of the iron-sulfur cluster is not obviously unreasonable, but the preliminary results of Johnson and Holm[36)] on acylation of synthetic 4 Fe–4 S systems suggest that substitution of thiolate by more electronegative oxygen ligands causes a *positive* rather than a negative shift in reduction potential. Clearly, however, the redox properties of phosphorylated iron-sulfur clusters are of potential interest.

## C. The Molybdenum-Iron Protein

It is generally agreed that the MoFe protein of nitrogenase from any of a variety of organisms has a molecular weight of ~220,000[26)], and is an $\alpha_2\beta_2$ tetramer with subunit molecular weights of ~50,000 and 60,000[37)]. Although a wide range has been reported[14, 26)] for both the iron content (from 12 to 32 atoms per molecule) and the molybdenum content (from 1 to 2 atoms per molecule), the most reliable estimate, based on chemical analysis of very highly purified preparations[38)] and analysis of the $^{57}$Fe Mössbauer spectra[39, 40)] appears to be 30 ± 2 Fe and 2 Mo per molecule. All workers agree that iron and sulfide are present in approximately equimolar quantities[26)], suggesting a value of ~30 $S^{=}$ per molecule; because of difficulties in the analysis of sulfide in these materials[41)], however, the correct number may be significantly different from this.

In contrast to those of the Fe protein, the metal components of the MoFe protein do not seem to consist solely of simple Fe-S tetramers or dimers familiar from the bacterial and chloroplast ferredoxins, respectively. This is clearly demonstrated by the EPR spectrum[42] at low temperature (6–20 K) of the MoFe protein as isolated, which shows a nearly axial signal with two g values near g = 4 (4.3 and 3.65) and one g value near g = 2 (2.01); double integration gives one spin per molybdenum. This spectrum is markedly different from that shown by any known Fe-S cluster, which typically have g-values near or below g = 2.0. Initial attempts to identify the species responsible for this signal utilized enzyme isolated from organisms grown separately on $^{95}$Mo (I = 5/2) and on $^{57}$Fe (I = 1/2); only $^{57}$Fe resulted in detectable broadening of the EPR spectrum, indicating that iron is involved in the chromophore[43]. Others[43, 44] suggested that the EPR spectrum was that expected for one of the two Kramer's doublets of an S = 3/2 system. An obvious candidate was Mo(III), a $d^3$ ion whose EPR spectrum was reported to have g values similar to those of the MoFe protein[45]. Recent work has shown, however, that typical Mo(III) complexes exhibit large $^{95}$Mo hyperfine interactions[46], ruling out an isolated Mo(III) complex as the source of the unusual EPR spectrum. Note, however, that the absence of observable $^{95}$Mo hyperfine interaction in the EPR spectrum of $^{95}$Mo-enriched MoFe protein does *not* preclude participation of Mo in the S = 3/2 species; preliminary results on the modulation of the EPR spin echo decay in $^{95}$Mo- and $^{96}$Mo-enriched MoFe protein indicate that Mo is indeed weakly coupled to the S = 3/2 center[47].

Combined EPR and $^{57}$Fe Mössbauer spectroscopic studies of the MoFe protein in various overall oxidation states[39, 40, 42, 44, 48, 49] have provided strong evidence for the presence of six metal clusters: two M centers that are the protein-bound form of the FeMo-cofactor (a novel Mo-Fe-S cluster) and four unusual tetranuclear iron clusters referred to as the P clusters. These will be discussed separately below. In addition, Mössbauer spectra of MoFe proteins from *Azotobacter vinelandii, Clostridium pasteurianum,* and *Klebsiella pneumoniae* all show an additional component termed S[39, 40, 42], which accounts for ~6% of the total Fe present (2 Fe per molecule) and which has Mössbauer parameters ($\Delta E_Q \simeq 1.35$ mm/s; $\delta = 0.60$ mm/s) different from those expected for likely impurities such as high-spin $Fe^{2+}$ or $Fe^{3+}$ ("adventitious" iron). At present, it is difficult to decide whether species S is an unusual and persistent impurity or an integral part of the MoFe protein.

### *1. The Iron-Molybdenum Cofactor (FeMo-co)*

In 1974, it was demonstrated[50] that a mutant strain (UW45) of *A. vinelandii* produced an inactive MoFe protein that could be activated by addition of acid extracts of native MoFe protein, suggesting that the UW 45 protein was lacking a soluble low molecular weight cofactor. Use of N-methylformamide to extract the MoFe protein permitted the isolation of this cofactor in relatively pure form[51]. This species was reported to contain Mo, Fe, and S in 1 : 8 : 6 molar ratios, and was termed the iron-molybdenum cofactor, or FeMo-co. FeMo-co is an extremely oxygen-sensitive material that is not stable in aqueous solution, and that is clearly distinct from the molybdenum cofactor that is apparently common to all other molybdenum enzymes[52, 53]. It has been reported[54] that isolated FeMo-co retains the ability to catalyze the reduction of acetylene, but not $N_2$, by borohydride, suggesting that it is the site of acetylene reduction in the intact protein. The fact that acetylene reduction activity exists in the presence of oxygen, which destroys the ability of FeMo-co to reconstitute mutant MoFe protein, implies that the observed acetylene reduction activity may simply reflect the general ability of molybdenum-thiol[55], molybdenum-cyanide[56], and iron-sulfur[57] systems to reduce acetylene.

*a. Composition.* Because of the extreme sensitivity of FeMo-co, it has been difficult to obtain definitive results on its chemical composition. The best recent analyses give a 1 : 7 : 4 ratio of Mo : Fe : S[38, 58, 59]; the low value for sulfide is supported by preliminary $^{35}S$ radioactive labeling experiments[60]. Ion exchange chromatography and electrophoresis[61] show that FeMo-co is definitely anionic. Assuming the lowest reasonable oxidation sates for the metals ($Fe^{2+}$, $Mo^{3+}$) and an oxidation state of $-2$ for sulfide, one would still predict a net charge of $+9$ for FeMo-co based on the analytical data. This clearly implies the existence of additional negatively charged constituents, yet the apparent molecular weight of FeMo-co determined by Sephadex G-100 chromatography in NMF is close to the sum of the metal and sulfur components[62]. Potential candidates in the original procedure for the missing negative charge include citrate, phosphate, dithionite, and sulfite. Recent work, however, has shown that none of these are essential components of FeMo-co, and preparations containing only sodium, chloride, and hydroxide ions, NMF, and water in addition to FeMo-co have been obtained[59]. This would seem to eliminate everything but hydroxide or oxide as additional anionic components of FeMo-co.

There has been a great deal of interest in the nature of the presumed endogenous ligands of FeMo-co; obvious possibilities include a peptide or amino acid, sugars, lipoic acid[63], and coenzyme A[63]. The recent report[64] on the extraction of an "MoFe cluster" from FeMo-co with acidified methylethylketone lends some support to the idea of an organic constitutent of FeMo-co, but careful analyses of purified FeMo-co have shown no evidence for amino acids[58, 61, 62], sugars[59], lipoate[59], or coenzyme A[59], and less than 0.5 phosphate per Mo[59]. There is thus no compelling evidence for an essential organic constituent of FeMo-co, and considerable circumstantial evidence against it. For example, FeMo-co apparently reacts with citrate, as shown by a significantly higher apparent molecular weight for FeMo-co preparations treated with citrate[60], with EDTA, as shown by disappearance of the FeMo-co EPR signal (*vide infra*)[65], and with thiophenol, as shown by a sharpening of the EPR spectrum[65, 66]. This suggests the presence of weakly bound, readily displaced ligands in FeMo-co; the prime candidate would appear to be NMF, since ~ 10 molecules of NMF per Mo cannot be removed from FeMo-co preparations even by lyophilization[59, 67].

*b. Spectroscopic Properties.* In general, the spectroscopic properties of FeMo-co are very similar to those of the parent MoFe protein, suggesting only relatively slight structural perturbations upon removal from the protein. Thus, the optical spectrum of FeMo-co shows only a gradually rising absorbance below 700 nm, and is bleached by oxygen or mercurials[51]. The EPR spectrum at temperatures below 30 K shows a broad spectrum with apparent g values of 4.6, 3.3, and 2.0[66], indicative of a slight increase in rhombic character. At higher temperatures ($> 12$ K) a peak at $g = 5.93$ is observed, due to population of the $\pm 3/2$ Kramers doublet[66]. FeMo-co is reversibly oxidized by methylene blue to an EPR-silent state[65], behavior analogous to that of the M centers in the MoFe protein upon thionine oxidation[39]. In contrast, attempts at reduction of isolated FeMo-co to the EPR-silent, integral spin state observed in the MoFe protein under nitrogen fixing conditions[42] have not been successful[65].

Low temperature Mössbauer spectra of isolated FeMo-co[66] are very similar to the magnetic spectrum assigned to the M centers of the MoFe protein[48], except for a slight broadening of the lines and for the presence of a quadrupole doublet ($\Delta E_Q \simeq 0.9$ mm/s, $\delta \simeq 0.4$ mm/s) due to a small amount (~20%) of iron present in another environment. Detailed analysis of the magnetic spectrum of the M centers in the MoFe protein has shown the presence of six magnetic subcomponents, each corresponding to approximately 1 Fe per Mo[48]; this provides independent evidence for the association of 6 Fe with each S = 3/2 M center in the MoFe protein and, by inference, in the FeMo-cofactor. Of the six magnetic subcomponents, three are characterized by a positive magnetic hyperfine coupling constant $A_O$, and three by a negative $A_O$, consistent with a spin-coupled structure[48].

Although detailed structural information on the MoFe protein and FeMo-co is not yet available from X-ray diffraction methods, the technique of X-ray absorption spectroscopy has now been used to probe the immediate environment of Mo in both the MoFe protein and FeMo-co and of Fe in FeMo-co. Because of the absence of any other spectroscopic response attributable to Mo, to date X-ray absorption spectroscopy has been employed chiefly to study the molybdenum site. Measurements of the Mo K absorption edge position were found to be consistent with sulfur coordination, and not particularly characteristic of the Mo oxidation state[68,69]. Analysis of the Mo K-edge extended X-ray absorption fine structure (EXAFS) spectra has given more detailed structural information, and confirmed the cluster nature of FeMo-co. The Mo EXAFS spectrum of the MoFe protein has been intrpreted as arising from scattering due to a set of 4 S atoms at 2.35 Å, 2–3 Fe atoms at 2.72 Å, and 1–2 additional S atms at ~2.47 Å from Mo[69–71]. There is no evidence for the presence of Mo=O, Mo=S, or Mo-Mo interactions[69,70]. Mo EXAFS spectra measured for the MoFe protein in an argon atmosphere are insignificantly different from those obtained in the presence of dinitrogen[72] suggesting that, in the resting state at least, no more than a small fraction of MoFe protein molecules, if any, contain $N_2$ directly bound to Mo. Removal of FeMo-co from the protein results in only minimal changes in the Mo-S and Mo-Fe interactions, demonstrating that the basic structure is conserved. The only obvious difference in the Mo EXAFS of FeMo-co compared to that of the parent protein is the appearance of a set of 2–3 low Z (O or N) scatterers at 2.10 Å from Mo[67]; these have been attributed to coordination of exogenous O or N donors added during isolation (e.g., NMF, citrate) to molybdenum.

Very recently, it has proven possible to obtain high quality Fe K-edge EXAFS data on frozen solutions of FeMo-co in NMF. Comparison of the Fourier transform of the data to those of the EXAFS spectra of two recently prepared synthetic Mo-Fe-S clusters with oxygen ligation to Fe (see Sect. III) shows clearly that the basic environment of iron is very similar in all three systems. Preliminary analysis of the FeMo-co data suggests 1.3 ± 1.0, O (or N) atoms at 1.8 Å, 3.4 ± 1.6 S atoms at 2.25 Å, 2.3 ± 0.9 Fe atoms at 2.66 Å, and 0.4 ± 0.1 Mo at 2.76 Å from Fe[73]; these results are clearly average values of parameters for at least two distinct iron sites.

*C. Proposed Structures.* That the FeMo-cofactor is a structurally novel cluster containing Fe, Mo, and $S^=$ is suggested by its composition and spectroscopic properties, by the observation that it is resistant to conditions that destroy (i.e., dilute acid[50,67] or 80% *(v/v)* 5 M imidazole in hexamethylphosphoramide[14]) or react with[66] normal protein-bound Fe-S clusters, and by the report that acid hydrolysis of the FeMo protein produces a mixture of tetrathiomolybdate, $MoS_4^{2-}$, and trithiomolybdate, $MoOS_3^{2-}$ [74]. Since tetrathiomolybdate cannot be formed under the conditions used in the acid hydrolysis experiment[75], this last finding together with the Mo EXAFS data strongly suggest a structure in which Mo is coordinated by four $S^=$ that presumably act as bridging ligands to Fe atoms. Within the constraints imposed by the body of data discussed above, several structures have been proposed for FeMo-co; some of them are shown schematically in Fig. 2. Although these are all to a greater or lesser degree speculative, they are worth

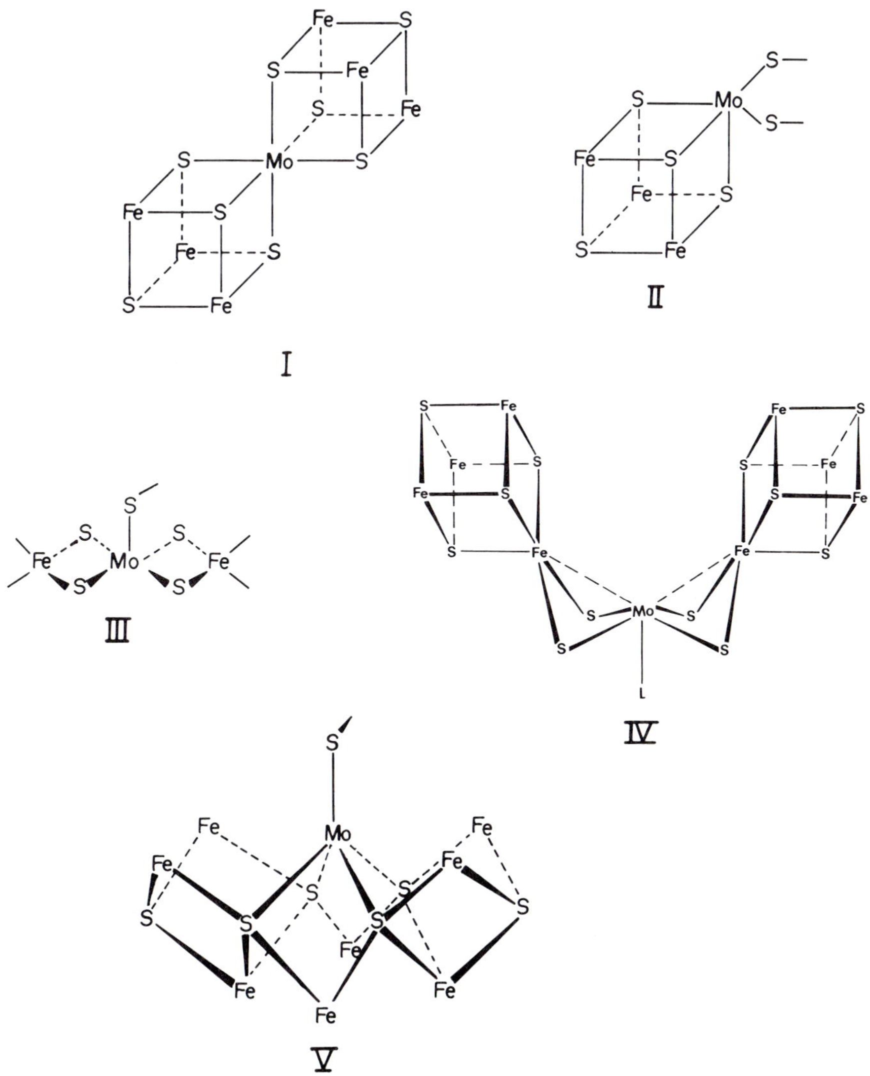

Fig. 2. Schematic of structural models proposed for the FeMo-cofactor

discussing briefly inasmuch as they provide a focus for synthetic efforts and illustrate some of the problems facing synthetic inorganic chemists. It is worth noting that all of these structures are based on a stoichiometry of one Mo per FeMo-co. Quantitation of the EPR[42)] of the MoFe protein gives one S = 3/2 center per Mo; this is consistent with either two FeMo-co units, each of which contains one Mo atom and one S = 3/2 center, or one FeMo-co unit, containing two Mo atoms and two independent S = 3/2 centers, per MoFe protein. Although firm evidence is lacking, the former alternative seems more likely based on the absence of detectable dipolar interactions between the S = 3/2 centers in the EPR spectrum, and on the observed molecular weight of 800–1500[62)], which is more consistent with a stoichiometry based on one Mo per FeMo-co.

Structure **I** shown in Fig. 2 consists of two cubane-like Fe clusters joined by a Mo atom at a common vertex. Although this model provides a chemically reasonable structure in good agreement with the spin-coupled $Fe_6Mo$ system indicated by Mössbauer measurements[48)], it has twice the reported sulfide content per Mo, and its predicted Mo K edge EXAFS spectrum[76)] is in poor agreement with that of the MoFe protein. Further, **I** contains Mo in a six-coordinate, octahedral site that appears to be coordinatively saturated, and which does not provide an obvious site for substrate binding and reduction.

Both structures **II** and **III** for the immediate environment of Mo in the FeMo protein were suggested as being consistent with the Mo EXAFS data[69)], but with the availability of EXAFS data on a synthetic complex with a very similar structure (see Sect. III), more recent data on the FeMo protein were analyzed exclusively in terms of the $MoFe_3$ cubane structure **II**[71)]. The major problem with this structure is that it contains only 3 Fe per Mo, and it is difficult to see how the structure can be further elaborated to include an additional 3–4 Fe without connecting at least one more Fe to the Mo via one or more sulfur bridges (i.e., tending toward structure **I**).

From the point of view of developing a chemically reasonable model for FeMo-co, linear structures such as **III** (with or without the fifth, axial sulfur ligand) seem much more promising, and are in at least as good agreement with the Mo EXAFS data as **II**[77)]. In practice, it is very difficult to distinguish between 2 and 3 Fe atoms as nearest neighbors to Mo in fitting the Mo EXAFS data; both cases give fits of approximately equal quality and result in virtually identical Mo-Fe and Mo-S distances[77)]. Virtues of structures related to **III** are the apparent coordinative unsaturation at the Mo atoms, providing an obvious site for substrate binding, the fact that a synthetic complex with a related structure is known (Sect. III), and the conceptual ease of elaboration to include a total of 6–8 Fe atoms per Mo. The two Fe atoms shown in **III** could each be part of either an Fe-S dimeric or tetrameric cluster[3–5)] or even one of the recently discovered 3 Fe clusters[78–80)]; due to the long (> 4 Å) distances of the additional Fe and/or S atoms from the Mo atoms, these would not be visible in the Mo EXAFS spectrum. An example of a structure based on **III** is **IV**, in which an $MoS_4$ unit with unspecified Mo oxidation state bridges two 4 Fe-4 S clusters. The major problem with a structures such as **IV**, in which the Fe atoms are part of a known $(Fe\text{-}S)_n$ (n = 2–4) center, is that these require the presence of significantly more sulfide per Mo (up to 12 $S^=$/Mo) than is found experimentally. A possibility that would solve this problem (and the problem of insufficient negative charge discussed earlier) is the presence of oxide- or hydroxide-bridged Fe clusters of unprecedented structure. Since this eventuality would radically alter the synthetic strategies used in developing models for FeMo-co (see Sect. III for a discussion of strategies employed to date), an accurate value of the $S^{=}$/Mo ratio in pure FeMo-co is of the utmost importance; application of alternative analytical methods is clearly desirable.

The final structure proposed to date for FeMo-co is the *nido*-cubane or "string-bag" cluster **V**[81]; it is a compromise between the cubane (**I, II**) and linear (**III, IV**) models. Attractive features of this model are the ease in accomodating Fe/Mo ratios from 6–8 by deletion of one or two Fe atoms, the potential for coordinative unsaturation of Mo, and the existence of structurally related inorganic complexes such as Roussin's black salt[82]. Problems with the model include the presence of six Fe rather than the two or three revealed by EXAFS as neighbors of Mo, and the presence of more than four $S^{=}$ per Mo. Although the latter can be easily overcome by substituting oxide for the sulfur atoms not attached to Mo, it is difficult to see how this structure could be consistent with the Mo EXAFS data.

## 2. *The P Clusters*

The presence of two FeMo-co units, each containing 6–8 Fe atoms, accounts for only 12–16 of the 30 ± 2 Fe atoms present in the FeMo-protein. Cluster displacement studies using fluorinated thiols and $^{19}$F NMR determination of Fe-S cluster[83,84], cluster transfer experiments using low molecular weight apoferredoxins to accept displaced Fe-S clusters followed by reduction and EPR quantitation[14], and quantitation of the EPR spectra of MoFe protein in organic solvents[14,47,66] all lead to the conclusion that four tetrameric iron-sulfur clusters can be removed from the MoFe protein under appropriate conditions, in addition to FeMo-co. This is consistent with earlier EPR studies of CO-inhibited nitrogenase, which detected an EPR signal due to a 4 Fe-4 S (or 2 Fe-2 S) center in the net 1+ oxidation state at low CO concentrations and an EPR signal typical of a 4 Fe-4 S center in the net 3+ oxidation state at higher CO concentrations[14,85].

In contrast, neither magnetic circular dichroism[86,87] nor $^{57}$Fe Mössbauer spectroscopy give results typical of 4 Fe-4 S clusters in any of the oxidation states (1+, 2+, 3+) known from simple Fe-S proteins. In addition to the magnetic spectrum attributable to the M centers (FeMo-co) and the minor quadrupole doublet referred to above as species S, Mössbauer spectra of MoFe protein from all three species examined in the resting (as isolated) oxidation state show two quadrupole doublets in a 3:1 intensity ratio[39,40,42,44,49]. In *A. vinelandii* MoFe protein, the less intense has parameters ($\Delta E_Q$ = 3.02 mm/s, $\delta$ = 0.69 mm/s[42]) typical of high spin ferrous iron in a tetrahedral sulfur environment (e.g., reduced rubredoxin[88]) and has been termed, appropriately enough, component $Fe^{2+}$ [42]; the more intense has $\Delta E_Q$ = 0.81 mm/s and $\delta$ = 0.64 mm/s, and has been termed component D[42]. Components D and $Fe^{2+}$ account for 40 ± 2 and 13 ± 0.5% of the total iron present, or 12 and 4 Fe atoms, respectively, per molecule[40]. Mössbauer measurements in strong magnetic fields showed that both component D and $Fe^{2+}$ are in a *diamagnetic* environment. The observed 3:1 stoichiometry and the necessity of explaining how an apparently high spin ferrous iron site could be effectively diamagnetic suggested that these two types of iron resided in tetranuclear clusters comprising one $Fe^{2+}$ and three D iron atoms each; the observed diamagnetism was ascribed to intramolecular antiferromagnetic coupling[42]. These tetranuclear units have been termed the P clusters[39].

Additional evidence for the existence of these novel spectroscopic entities comes from combined EPR and Mössbauer studies[39, 40] and low temperature MCD spectra[87] of MoFe proteins reversibly oxidized with the dye thionine. Oxidation of MoFe protein from *A. vinelandii* proceeds in two phases: removal of up to four electrons per molecule causes no change in the g = 4.3, 3.65, 2.01 EPR signal due to the M centers, but removal of two more electrons per molecule causes complete loss of the EPR spectrum. Parallel Mössbauer experiments show that removal of four electrons causes complete loss of the D and $Fe^{2+}$ spectral components, with concomitant appearance of a new magnetic spectrum corresponding in intensity to 16 Fe atoms per molecule[39, 40]. The only way in which four electrons can result in 16 Fe atoms going from a diamagnetic to a paramagnetic state is if they are coupled to form clusters. Analysis of this magnetic spectrum in large applied magnetic fields shows the presence of four spectral subcomponents, three with negative and one with positive nuclear hyperfine coupling constants, providing additional evidence for a spin coupled structure[40]. It should be noted, however, that the idea of four equivalent P clusters is not universally accepted[49]. No EPR spectrum corresponding to the oxidized P clusters has yet been reported; this has been attributed to a combination of low intensity due to the large anisotropy expected for an $S \geq 3/2$ ground state with large negative zero-field splitting and relaxation broadening[39]. Low temperature MCD spectra of thionine-oxidized MoFe protein from *K. pneumoniae* are dominated by an unusual spectrum assigned to the oxidized P clusters[87], and the MCD magnetization curves are consistent with either an $S = 5/2$ or 7/2 unit. Room temperature magnetic susceptibility measurements on thionine-oxidized MoFe protein from *A. vinelandii* are most consistent with an $S = 5/2$ spin state for the oxidized P clusters[89].

If the P clusters are variants of "normal" 4Fe-4S clusters, then there are three possible choices of oxidation state for the P clusters in the resting enzyme (diamagnetic $P^N$): the $2^+$ state found in oxidized bacterial ferredoxins and reduced high potential iron proteins[3–5], the all-ferric (net charge 4+), or the all-ferrous (net charge O) oxidation states. The second of these has never been detected in proteins or even electrochemically in model compounds[3, 4], while the last can be generated electrochemically[3, 4] or chemically[90] with model systems. The $4^+$ oxidation state seems to be ruled out by the large average isomer shift (*ca.* 0.65 mm/s) and its expected instability, especially in contact with sodium dithionite, while the characteristic near-infrared MCD spectrum of the 2+ oxidation state has not been observed[86, 87]. Together with the highly reducing nature of the system, these considerations strongly suggest that the diamagnetic $P^N$ clusters are some variant of 4Fe–4S clusters in the all-ferrous oxidation state.

How then to explain the existence of two separate Mössbauer components? Based on the resemblance of component $Fe^{2+}$ to the tetrahedral sulfur site of reduced rubredoxin, it seems likely that it too is ligated by four sulfur atoms, suggesting normal cysteinyl mercaptide coordination to at least one corner of a 4Fe-4S cluster. The most obvious way for the protein to differentiate the other three iron atoms would seem to be by the use of non-thiolate ligands, although geometric distortion from tetrahedral coordination or addition of a fifth ligand to each site cannot be ruled out, nor can distortion of the 4Fe-4S core. Based on the realtively low affinity of ferrous iron for saturated nitrogen ligands[91], the most likely alternative to cysteine is oxygen ligation, either by phenolate or carboxylate residues. Hypothetical models for the P clusters with tyrosyl phenoxide and glutamyl carboxylate ligands are shown in Fig. 3; there is no evidence favoring tyrosine over glutamate or aspartate side chains.

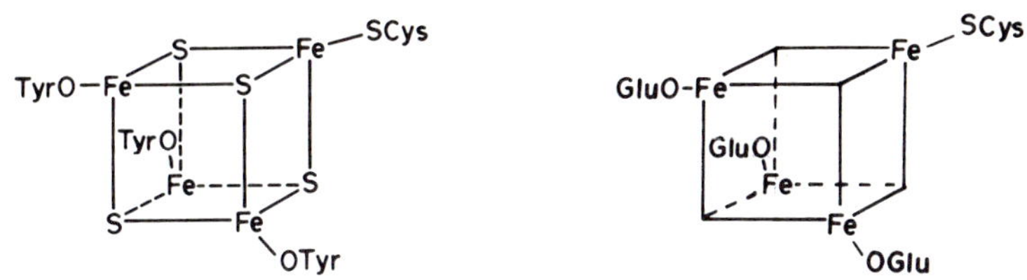


**Fig. 3.** Possible models for the P cluster involving oxygen ligation at three vertices of a 4 Fe-4 S core. The right hand structure could equally well be drawn with aspartate rather than glutamate

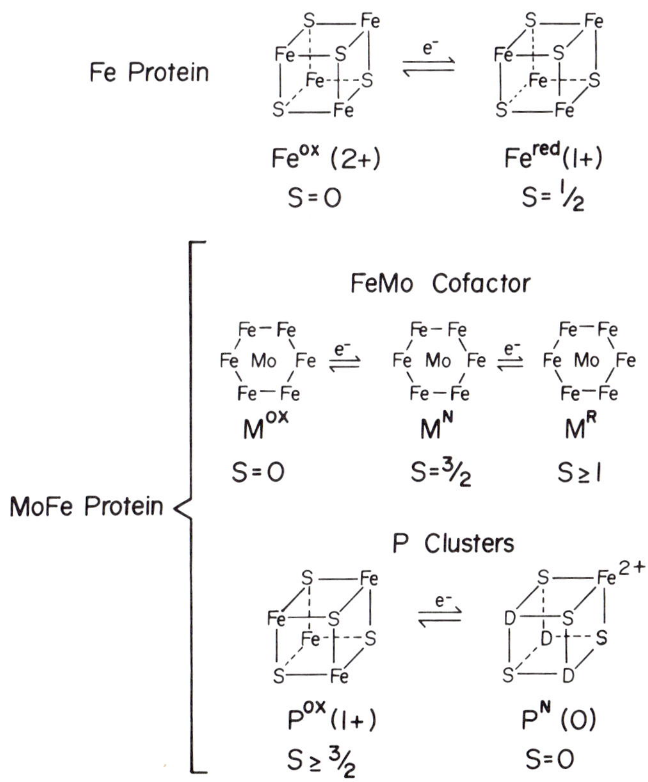


**Fig. 4.** Summary of the known oxidation states and magnetic properties of the metal-sulfur centers of the nitrogenase system. The numbers in parentheses indicate the presumed overall charge on the $Fe_4S_4$ core

The properties of known oxidation states of the prosthetic groups of the nitrogenase proteins are summarized in Fig. 4. The subsequent section will consider the progress made by synthetic inorganic chemists in developing models for these unusual metal clusters.

# III. Synthetic Mo-Fe-S Clusters as Models for the FeMo-Cofactor

## A. Rationale

The information presented above (Sect. II.C.1) demonstrates that the FeMo-cofactor of nitrogenase consists of a novel prosthetic group, a polynuclear Mo-Fe-S cluster. It is unlikely that the occurrence of this unusual species in the enzyme that accomplishes the difficult task of reducing dinitrogen to ammonia is coincidental; that is, the feasibility of the enzymatic reaction must be due at least in part to specific chemical properties of the cofactor, appropriately modulated by the protein environment. Unfortunately, very little is known about the properties of such polynuclear units, as the first examples were reported only in 1978[71, 92]. Thus, even when the structure of the FeMo-cofactor is determined (as it certainly will be, despite the formidable problems involved in crystallizing very small quantities of extremely oxygen-sensitive materials), it will be difficult to assess potentially novel features in the absence of a substantial body of information on the fundamental properties of such units.

Further, it is probable that an analog of the FeMo-cofactor will be synthetically accessible. This assertion is based on the previous work on the synthesis of polynuclear metal species, expecially the tetrameric and dimeric iron-sulfur clusters that form the prosthetic groups of the low molecular weight iron-sulfur proteins[3–5]. At present, it in general does not seem possible to set out to deliberately synthesize polynuclear clusters of a particular composition and degree of oligomerization; rather, *they assemble themselves from a mixture of suitable components* because they are the most stable species potentially present. The unusual stability of the FeMo-cofactor toward a variety of reagents (which may reflect the high stability of Mo-S bonds and highly condensed iron-sulfur (or iron-oxygen?) polyhedra) together with biosynthetic considerations regarding assembly of such a unit within an organism, suggest that the same will be true of the basic inorganic structural unit of the FeMo-cofactor. This system is considerably more complicated than those which yield the simple iron-sulfur clusters, inasmuch as the appropriate form of molybdenum, the identity of all the non-metallic cluster atoms, and the nature (or even the existence) of the organic component are all uncertain at present.

A particularly significant limitation has been the availability of suitable soluble molybdenum-containing starting materials. All characterized complexes reported to date have used tetrathiomolybdate, $MoS_4^{2-}$, which is well known to coordinate to other metal ions[93], as a source of molybdenum despite its dubious biosynthetic suitability. (The only exception to this is the complex $(\eta^5\text{-}C_5H_5)_2Mo(S\text{-}n\text{-}C_4H_9)_2FeCl_2$[94, 95]; the presence of bridging thiolate rather than sulfide and the obviously abiological cyclopentadienyl groups make this compound of limited biological relevance.) Tetrathiomolybdate is so far unique in providing a soluble, reactive mononuclear source of molybdenum (and sulfur) that readily forms clusters with iron salts. Other binary Mo-S complexes include the $MoS_9^{2-}$ ion[96], produced by reaction of sulfur with tetrathiomolybdate, and the series of polynuclear Mo-S cluster anions reported by Müller and coworkers[97, 98]. Although these compounds are of interest in their own right, it is not clear that they will provide pathways to new types of Mo-Fe-S clusters. Use of obvious precursors, such as the penta-, tetra-, and trivalent halides, seems to be limited by the formation of relatively unreactive dimers, such as $[Mo_2(SR)]^{3-}$[99], $[Mo_2S_4(SCH_2CH_2S)_2]^{2-}$[100, 101], and $[Mo_2S_4(SR)_4]^{2-}$[102]. The only exceptions to this appear to be $Mo(SCH_2CH_2SCH_2CH_2S)_2$[103] and the coordinatively unsaturated monomer, Mo(S-*t*-Bu)$_4$[104]; the latter is highly reactive[105], and appears especially promising. Finally, although there exists a vast number of soluble molybdenum complexes containing the molybdenyl (Mo=O) moiety[2], researchers have apparently been deterred from examining them as starting materials because of the extreme stability of the Mo=O bond and the consequent difficulty in cleaving it[2].

Two basic types of Mo-Fe-S cluster have now been characterized, containing either the $MoFe_3S_4$ cubane core or a more extended, "linear" array based on $MoS_2Fe$ units. The preparation, structures, properties, and reactivity of each of these units will be described, followed by a brief discussion of attempts at preparation of "string-bag" clusters. In addition to comparison of essential features to available data on the FeMo-cofactor, the properties of tungsten analogs, where available, will be compared to those of the Mo-Fe-S clusters, in order to ascertain why tungsten is apparently not able to replace molybdenum in nitrogenase[106].

## *B. $MoFe_3S_4$ Cubane Clusters*

### *1. Synthesis*

The reaction of $FeCl_3$ with alkyl or aryl thiolates and tetrathiometallates in alcoholic solvents has so far resulted in isolation of salts of five distinct types of anionic clusters (M = Mo or W): $[M_2Fe_6S_9(SR)_8]^{3-}$, **1**[71, 76, 107]; $[M_2Fe_6S_8(SR)_9]^{3-}$, **2**[76, 92, 108–111]; $[M_2Fe_6S_8(SR)_6(OR')_3]^{3-}$, **3**[111, 112]; $[M_2Fe_7S_8(SR)_{12}]^{3-}$, **4**[107, 108, 113]; and $[M_2Fe_7S_8(SR)_{12}]^{4-}$, **5**[107, 108, 113]. (Henceforth, **1–5**, etc., will refer to complexes with M = Mo, and **1-W**, etc., to complexes with M = W.) Isolation of a particular product depends upon the nature of the substituent R, choice of M = Mo or W, solvent, reaction time, cation, and ratio of thiolate and iron to tetrathiometallate[107, 108, 111, 112].

All of these products of self assembly reactions contain two $MFe_3S_4$ cores connected by a variety of bridging groups; their structures are shown schematically in Fig. 5. Systems **1**, **2**, and **3** differ only in the nature of the three groups connecting the molybdenum or tungsten atoms, having one sulfide and two mercaptide, three mercaptide, and three alkoxide bridging ligands, respectively. Systems **4** and **5** differ in having an additional ferric or ferrous iron atom and three mercaptide ligands inserted between the $MFe_3S_4$ cubes of **2**. Only in **1** are the four sulfide ligands of the starting tetrathiometallate retained, one being replaced by thiolate or alkoxide in **2–5**. These constitute the only products obtained by direct reaction of simple constituents; others that contain, for example, halide substituents in place of the terminal thiolates of **2** or a single $MoFe_3S_4$ core have now been prepared using **2** or **4** as starting materials and will be discussed in Sect. III.B.4.

### *2. Structures*

The structures of some eight Mo-Fe-S double cubane clusters of the types shown in Fig. 5 have now been determined by single crystal X-ray diffraction studies. Known structures include one example of **1** (R = Et[71, 76]), four examples of **2** (R = Et[76, 110]; R = Ph[109];

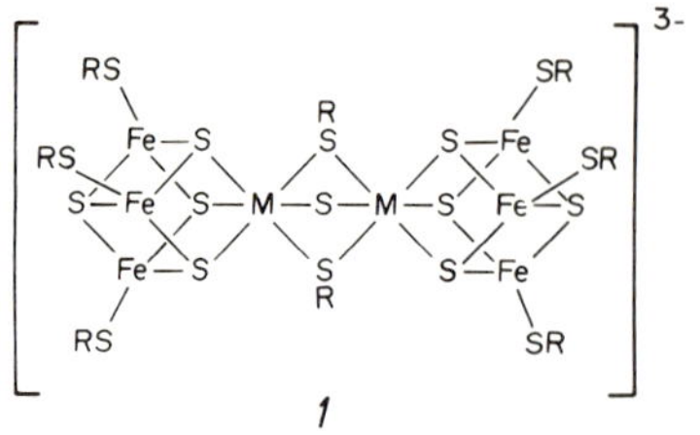

*1*

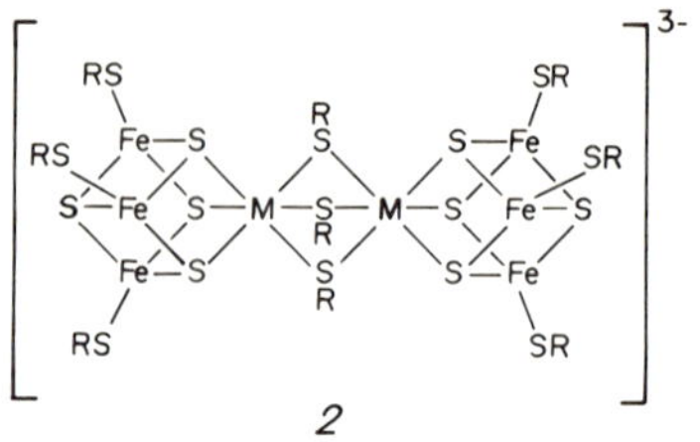

*2*

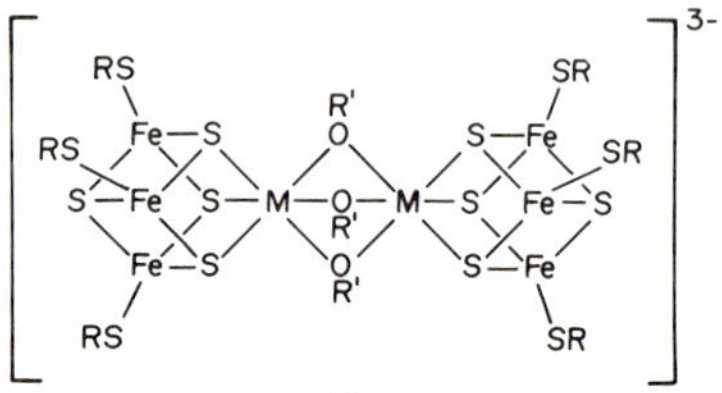

*3*

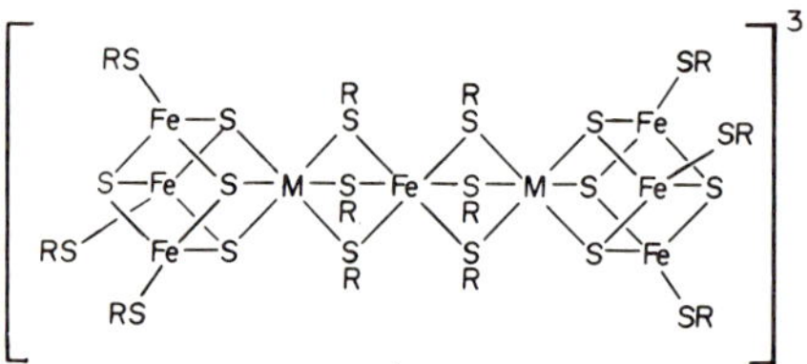

*4*

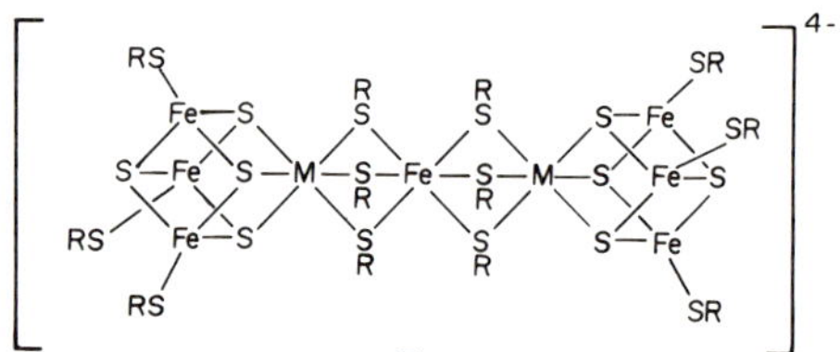

*5*

**Fig. 5.** Schematic of the structurally characterized $MoFe_3S_4$ double cubane clusters prepared to date

R = $CH_2CH_2OH$[114]), one example of **3** (R = Ph, R′ = Me[115])), one example of **4** (R = Et[113])), and one example of **5** (R = $CH_2Ph$[113])). Typical structural parameters for each type are summarized in Table 1 along with comparable data for species containing $Fe_4S_4$ cores. Inspection of Table 1 and the published data (esp. Ref. 76, 113) reveals a number of structural features common to the $MoFe_3S_4^*$ cubane cores of **1**–**5**. All are comprised of two distorted, interpenetrating tetrahedra containing $MoFe_3$ and $S_4^*$ vertices; the latter is significantly larger than the former. Consequently, each face of the cluster is a distinctly nonplanar rhomb, while the planes containing body diagonals are close to perfect. Many, but not all, of the structures determined possess a crystallographically imposed threefold axis containing the Mo atoms and opposite sulfides within the $MoFe_3S_4^*$ cube. Nonetheless, all exhibit a dominant threefold distortion of each $MoFe_3S_4^*$ core, which takes the form of a *ca.* 0.1 Å compression along the Mo-S* body diagonal.

The Mo-Fe distances of ca. 2.73 Å are very close to those reported for the Fe-Fe distance in the simple $[Fe_4S_4(SR)_4]^{2-,3-}$ clusters[3–5, 118, 120]), indicating that a second-row transition element can be accomodated in a cubane structure with minimal structural perturbation. The observed Mo-Fe distance is virtually identical to that found by EXAFS studies for the Mo site of nitrogenase[69–71]). The Fe-S and Fe-S* distances are slightly ($\leq 0.06$ Å) but significantly shorter than corresponding distances in the $Fe_4S_4^*$ units, suggesting a higher ($> +2.5$) average Fe valence. Most importantly, the Mo-S* distances of *ca.* 2.35 Å are almost 0.2 Å longer than those in the parent $[MoS_4]^{2-}$ ion[121, 122]) and

**Table 1.** Selected average parameters for $MoFe_3S_4^*$ clusters and related units[a]

| | **1**-Mo(R = Et)[b] | **4**-Mo(R = Et)[c] | **9**[d] | **1**-W(R = Et)[e,f] | **3**-W(R = Ph, R′ = Me)[f,g] | $[Fe_4S_4(SCH_2Ph)_4]^{2-}$ [h] |
|---|---|---|---|---|---|---|
| Mo-Fe | 2.723(2) | 2.730(14) | 2.725(29) | 2.718(3) | 2.710(2) | |
| Mo-Fe′ | | 3.319(1) | 3.092(3) | | | |
| Mo-Mo | 3.668(4) | 6.638(2) | | 3.674(3) | 3.174(2) | |
| Fe-Fe | 2.687(3) | 2.712(17) | 2.709(19) | 2.702(5) | 2.693(4) | 2.746 |
| Mo-S* | 2.351(3) | 2.357(7) | 2.338(16) | 2.343(5) | 2.335(4) | |
| Mo-S(R) | 2.567(4) | 2.559(8) | 2.150(24)[i] | 2.571(5) | 2.123(9)[i] | |
| Fe-S* | 2.254(4) | 2.264(4) | 2.255(24) | 2.275(6) | 2.275(4) | 2.286 |
| Fe-S*′ | 2.273(5) | 2.282(4) | 2.269(15) | 2.272(6) | 2.278(4) | |
| Fe-S(R) | 2.232(5) | 2.256(4) | 2.234(19) | 2.234(7) | 2.245(5) | 2.251 |
| S*-Mo-S* | 102.47(11) | 102.7 | 102.7 | 103.6(1) | 103.9(1) | |
| S*-Fe-S* | 108.86(18) | 108.9 | 108.2 | 108.0(2) | 107.8(2) | 104.1 |
| S*-Fe-S* | 104.17(13) | 103.8 | 103.7 | 104.3(2) | 104.6(1) | |
| Mo-S*-Fe | 72.47(12) | 72.4 | 72.8 | 72.1(2) | 71.9(1) | |
| Fe-S*-Fe | 73.17(14) | 73.6 | 73.8 | 72.8(2) | 72.6(2) | 73.81 |
| Fe-S*′Fe | 72.47(18) | 72.9 | 73.3 | 73.0(2) | 72.8(2) | |
| Mo-S(R)-M | 91.20(17) | 85.8 | 93.7[i] | [j] | 96.7(5)[i] | |

[a] Distances in Å, angles in deg. Fe′ denotes Fe not in $MoFe_3S_4^*$ core. S* denotes S within $MoFe_3S_4$ core bridging Fe and Mo; S*′ denotes unique S within $MoFe_3S_4$ core bridging 3 Fe. S(R) denotes mercaptide sulfur

[b] Ref.[76)]

[c] Ref.[113)]

[d] Ref.[116)]

[e] Ref.[117)]

[f] Values listed under Mo are for W analog

[g] Ref.[112)]

[h] Ref.[118)]

[i] S(R) is replaced by O(R)

[j] Not reported

essentially identical to those found by EXAFS studies of the MoFe protein. This longer distance almost certainly indicates that the molybdenum has formally been reduced from +6 to some lower oxidation state, with a concomitant increase in the radius of the molybdenum atom and a substantial decrease in multiple bond character in the Mo-S* bond.

The triply bridged binuclear $ML_3M$ unit that comprises the central portion of clusters **1–3** is not uncommon in Group VI chemistry (*cf.* $[Mo_2Cl_q]^{3-}$ [123]). The Mo-Mo distances found to date (**1**, 3.19–3.31 Å[76]; **2**, 3.64–3.69 Å[76, 109, 110, 114]; **3**, 3.14 Å[115]) are, however, significantly longer than those found in other triply bridged binuclear Mo complexes (ca. 2.6–2.7 A) and are consistent with the absence of any direct Mo-Mo interaction. In contrast, the "extended" triple bridges of clusters **4** and **5** are unique[113]. In these systems, the Mo-Fe (bridging) distances (**4**, 3.32 Å; **5**, 3.46 Å) are also consistent with the absence of direct metal-metal interactions. In all cases examined, the Mo-S (bridging) distances are ~2.56 Å, about 0.2 Å longer than the Mo-S(core) distances. The dimensions of the bridging distorted trigonal prismatic $Fe(SR)_6$ units in **4** and **5** have been interpreted in terms of the central Fe atom existing as low-spin Fe(III) in **4** and high-spin Fe(II) in **5**[113].

## 3. *Properties*

*a. Electronic Spectra.* The optical spectra of complexes **1**, **2**, **4**, **5**, where $RS^-$ is an alkyl thiolate, consist of a relatively featureless rising absorption throughout the visible region into the ultraviolet, superimposed upon which distinct bands at 389–396 nm and 265–290 nm (depending on R) are visible[76, 108, 110, 114]. Molar extinction coefficients vary from 30–40,000 $cm^{-1}$ (**1**, **2**) or 50–65,000 $cm^{-1}$ (**4**, **5**) for the lower energy band to 50–60,000 $cm^{-1}$ (**1**, **2**) or 80–100,000 $cm^{-1}$ (**4**, **5**) for the higher energy band[76, 108, 114]. Use of aryl thiolates shifts the spectral features for **2** to ~ 350 and 450 nm[92]. No spectral data have yet been reported for complexes of type **3** (M = Mo).

There are thus significant similarities between the spectra of these $MoFe_3S_4$-containing species and those of the $[Fe_4S_4(SR)_4]^{2-}$ clusters with similar terminal ligands[4, 124]. The major differences are a slight (20–30 nm) blue shift of the two major spectral features, and slightly poorer resolution of these features from the underlying absorption. The similarities extend to the solvent dependence of the spectra as well: blue shifts of ~20 nm and ~10 nm are observed for the first and second bands upon going from aprotic (DMSO) to aqueous solution[114], shifts comparable in magnitude to those observed for $[Fe_4S_4(SR)_4]^{2-}$ [125]. Taken together, these observations suggest a similar origin for the spectra in both cases, namely, mercaptide to core iron charge transfer transitions[126].

As a final point, the optical spectrum reported for FeMo-co[51] consists of a featureless steadily rising absorbance throughout the visible and ultraviolet regions, and is thus quite similar to the spectra of complexes **1**, **2**, **3**, **5**. The major difference is the lack of distinct bands in the FeMo-co spectrum, which may be attributable to the lack of thiolate in FeMo-co preparations[59] and to the effects of the NMF solvent on the optical spectra of known Fe-S systems[127].

*b. Proton Magnetic Resonance Spectra.* Because of their lack of detail and relative insensitivity to the particular cluster type, the optical spectra discussed above are not very useful for identification of compounds or for assessment of purity. Inasmuch as the synthetic schemes developed to date produce a single product only under very carefully defined conditions, an alternative spectroscopic means of identification is necessary. Proton NMR spectra have been reported for examples of clusters **1**, **2**, **4**, **5**, and are highly characteristic[108, 110, 111)].

Separate sets of resonances are observed for terminal and bridging thiolates in all cases, suggesting that all bridging and terminal ligands are equivalent and that exchange of thiolates between the two positions is slow on the NMR time scale. All spectra are characterized by the presence of isotropic shifts, due to the paramagnetic nature of the materials (see below), and are consequently highly temperature dependent. At a given temperature, however, the patterns and magnitudes of the isotropic shifts are consistent with the structures observed in the solid state. Thus, at room temperature all complexes containing aliphatic primary thiolates exhibit resonances far downfield of TMS (− 55 to − 64 ppm) that are assignable to Fe-$SCH_2$ protons of terminal thiolates[108, 110, 111)]. The position of the α-$CH_2$ protons of the bridging thiolates depends strongly on the cluster type: **1**, −17 to − 18 ppm; **2**, +3 to 4 ppm; **4**, − 15 to − 25 ppm (2 peaks); **5** − 80 to − 90 ppm (broad)[108, 110, 111)]. Peaks due to protons further from the paramagnetic center exhibit smaller isotropic shifts, as expected. Examination of the temperature dependence of the spectra of complexes **4** and **5** has shown that the presence of multiple or broad peaks due to bridging thiolates can be attributed to the intrinsic diastereotopism of these protons rather than to the presence of multiple geometric isoners[108)].

The spectra of **2** (R = Ph) and other arylthiolate analogs also exhibit separate sets of resonances for each type of ligand. At room temperature, peaks at − 13 to − 14 ppm (to low field of TMS) and at + 4 to + 5 ppm are assigned to the *meta* and *para* aromatic protons of the terminal thiolates, respectively, while peaks at − 8 to − 9, − 5.5 to − 6, and − 5 to − 5.5 ppm are assigned to the *ortho, meta,* and *para* protons of the bridging thiolates[111)]. These assignments have been confirmed by studies with *p*-Cl and *p*-$CH_3$ substituted ligands. The *ortho* proton resonance of the terminal thiolates is apparently too broad to be observed at room temperature, but has been observed above 80 °C and is near the *para* proton resonance (i.e., upfield of TMS).

These observations suggest substantially different origins for the observed isotropic shifts for terminal and bridging thiolate ligands. In particular, the behavior of the terminal thiolate protons parallels that observed for $[Fe_4S_4(SR)_4]^{2-,3-}$ [128, 129)]. The alternation of sign of *o*-H, *m*-H, *p*-H in **2** (R = Ph)[111)], the change in sign upon replacing *p*-H by *p*-$CH_3$, observation of negative methylene and methyl proton shifts for R = Et, the ratios of the shifts observed at different positions within a ligand, and the negative temperature coefficient of the isotropic shifts, which parallel the magnetic susceptibility, are all consistent with dominant contact interactions arising from terminal ligand to iron antiparallel spin transfer[108, 111)]. In contrast, the behavior of the isotropic shifts of the bridging ligands is more variable; bridging aromatic thiolates do *not* exhibit alternation of signs of shifts, and the bridging $SCH_2$ of **1** and **2** (R = Et) exhibit shifts of opposite sign, as do the $SCH_2$ and $CH_3$ protons of **4** and **5** (R = Et). These data are most consistent with the presence of a significant dipolar contribution to the observed isotropic shifts for the bridging thiolates[108, 111)].

*c. Mössbauer Spectra.* In an attempt to infer the approximate oxidation level of the Mo atoms and to assess the electron distribution with the $MoFe_3S_4$ units, $^{57}$Fe Mössbauer spectra of complexes of types **1**, **2**, **4**, and **5** have been measured. In zero applied field, all such complexes yield an asymmetric quadrupole doublet[76, 130], suggesting that the Fe atoms are not strictly equivalent. Complexes of type **4** and **5** exhibit an additional, less intense quadrupole doublet (intensity ~ 1/6 of the major doublet) that is clearly attributable to the bridging Fe atom[108]. The isomer shift and quadrupole splitting of these doublets is consistent with the presence of low-spin $Fe^{3+}$ and high-spin $Fe^{2+}$ in the bridging units of **4** and **5**, respectively.

The asymmetric doublet due to the $MoFe_3S_4$ units has been described as consisting of two[76] and three[130] unresolved quadrupole doublets with similar parameters. The average isomer shifts and quadrupole splittings at 77 K range from 0.27–0.33 mm/s (referenced to Fe metal at 77 K) and 0.95–1.20 mm/s, respectively. Previous work[131] has shown that the isomer shift of Fe in tetrahedral sulfur environments is linearly correlated with formal oxidation state. Using known Fe-S complexes and Fe-S phases as standards, the observed isomer shifts for the $MoFe_3S_4$ units correspond to mean oxidation states of Fe of + 2.5[130] to + 2.67[76]. In particular, the isomer shifts are slightly smaller than those observed for $[Fe_4S_4(SR)_4]^{2-}$ [132], which contain Fe in an average oxidation of + 2.5, and are thus consistent with slightly greater positive charge on the Fe atoms. An average oxidation state of + 2.67 would correspond to 2 Fe(III) and 1 Fe(II); assuming normal charges for bridging $S^{2-}$ and for $RS^-$, this leaves a formal valence of Mo(III) for both Mo atoms in **2**, **4**, and **5**[76]. Complex **1** is then best formulated as a mixed valence Mo(III)-Mo(IV) system, since the isomer shift of Fe is virtually identical in **1**, **2**, **4**, and **5**. The isomer shifts and quadrupole splittings observed for these systems are quite close to those reported for FeMo-co (0.25 and 0.81 mm/s, respectively)[48, 66], suggesting an average oxidation state of iron similar to that in the $MoFe_3S_4$ clusters.

Examination of the Mössbauer spectra of compounds **1**, **2**, **4**, and **5** in applied longitudinal magnetic fields of up to 80 kOe at 4.2 K has indicated the presence of two magnetic subsites in the $MoFe_3S_4$ cores in approximately 2 : 1 intensity ratios and with negative and positive magnetic hyperfine coupling constants, respectively[108]. These results show that the $MoFe_3S_4$ units possess paramagnetic ground states with appreciable unpaired electron density at all Fe atoms and intracluster antiferromagnetic spin coupling, a situation similar to that in $[Fe_4S_4(SR)_4]^{3-}$ [119, 133] and FeMo-co[48].

*d. Magnetic susceptibility*. Further evidence for the spin-coupled nature of the $MoFe_3S_4$ units is provided by the limited magnetic data available[71, 92, 108, 115, 134]. Compounds of type **1** clearly differ from **2** and **3** by one electron; these must be odd- and even-electron systems, respectively. Magnetic moments per $Fe_6Mo_2$ unit have been reported ($\mu(\mu_B)$, T (K)) for $(Et_4N)_3[Fe_6Mo_2S_9(SEt)_8]$ (4.87, 4.3; 5.67, 100; 5.83, 299) and $(Et_3NCH_2Ph)_3[Fe_6Mo_2S_8(SEt)_9]$ (5.62, 4.3; 5.54, 100; 5.73, 299)[108]. Thus, compounds **1** appear to have a significantly greater temperature dependence of the magnetic moment than **2**. Recently, complete $\chi(T)$ data have been reported for complexes of type **3** $(Et_4N)_3[Fe_6Mo_2S_8(SPh)_6(OMe)_3]$[134] and type **2** $(Bu_4N)_3[Fe_6Mo_2S_8(SPh)_q]$[115], together with a relatively simple spin-coupling model[134]. The basic assumption underlying this model is that the unpaired spin density is located entirely on the iron atoms. If oxidation states are assigned as 2 Fe(III), 1 Fe(II), and 1 Mo(III), as discussed above, and it is assumed that unpaired spins on the Mo atoms are effectively paired by a superexchange interaction via the bridging ligands, then a relatively simple model for the magnetic properties of the isolated $MoFe_3S_4$ cores can be utilized. Both Fe(III) centers are allowed to interact in an antiferromagnetic fashion to give a resultant spin, $S^\dagger$, which is then coupled to the S = 2

Fe(II) center to give a set of spin levels, S′, for the $MoFe_3S_4$ unit. Straightforward calculation of the susceptibility gave excellent fits to the data for **3** with exchange coupling of ~ − 180 $cm^{-1}$ for the Fe(III)-Fe(III) pair and − 60 $cm^{-1}$ for the $S^{\dagger}$-Fe(II) interaction[135]. These results are in reasonable agreement with magnetic studies of $(Ph_4As)_2[Fe_2S_2(S_2\text{-}o\text{-xyl})_2]$[135] and reduced 2 Fe ferredoxins[136], and with calculations on $[Fe_2S_2(SH)_4]^{3-}$[137]. The ground state of each cube is found to be S′ = 1, with essentially zero interaction between the two S′ = 1 units of complex **3**. The data for complexes **2**, while significantly different in appearance, can be fit using the same model with a weak intercluster coupling[115].

*e. Electron paramagnetic resonance spectra.* Preliminary reports suggest that the double cubane clusters **1–3** exhibit complex, highly temperature dependent EPR spectra. Thus, $(Et_4N)_3[Fe_6Mo_2S_9(SEt)_8]$ in frozen DMF solution shows apparent g values of 10, 4, 1.9, and 1.2[71], while solid $(Et_4N)_3[Fe_6Mo_2S_8(SPh)_6(OMe)_3]$ has apparent g values of 14.6, 6.3, 4.6, and 2.6[115]. The latter spectrum is not visible above 20 K, and it remains to be shown that it is consistent with the S′ = 1 ground state predicted from the magnetic studies[134]. Certainly to date no isolated systems containing $MoFe_3S_4$ cores even approximate the characteristic EPR spectrum of FeMo-co.

## 4. *Reactivity*

*a. Oxidation-Reduction Reactions.* The results of extensive electrochemical studies on complexes of types **1**[108], **2**[108, 115, 130], **3**[115, 130], **4**[108], and **5**[108] have been reported. As expected based on the rich electrochemistry exhibited by the $[Fe_4S_4(SR)_4]^{n-}$ clusters[3, 4, 124], the $MoFe_3S_4$-containing species are also capable of existing in several distinct-oxidation states. Most extensively studied are complexes of type **2**, the results for which may be summarized as follows:

*(i)* The complexes contain a pair of weakly coupled $MoFe_3S_4$ clusters, each of which is capable of undergoing an irreversible one-electron oxidation and two well-separated one-electron reductions; they thus exhibit a pair of one-electron oxidations and two pairs of one-electron reductions.

*(ii)* The first reduction potential for $[Mo_2Fe_6S_8(SR)_9]^{3-}$ (**2**) is generally within *ca.* 50 mV of the first reduction potential of the corresponding $[Fe_4S_4(SR)_4]^{2-}$ cluster.

*(iii)* The variation in first reduction potential with changes in the terminal ligands is also similar in the two cases, while variation of the bridging ligands has little effect.

*(iv)* The separation of the components of each pair of reductions or oxidations is relatively constant at ~200 mV, and is due to a weak coupling between the two $MoFe_3S_4$ clusters via the bridging unit.

*(v)* Only the first and second reduction processes appear to be strictly reversible electrochemically. Thus, the following three-membered electron-transfer series appears to contain the only stable, potentially accessible chemical species in these systems:

$$[Mo_2Fe_6S_8(SR)_q]^{3-} \underset{}{\overset{e^-}{\rightleftarrows}} [Mo_2Fe_6S_8(SR)_q]^{4-} \underset{}{\overset{e^-}{\rightleftarrows}} [Mo_2Fe_6S_8(SR)_q]^{5-} .$$

Indeed, successful isolation of the two-electron reduced product $[Mo_2Fe_6S_8(SPh)_q]^{5-}$ has recently been reported[138]; preliminary Mössbauer measurements show an increase in the

$^{57}Fe$ chemical shift (from 0.32 to 0.41 mm/s), consistent with substantial localization of the extra negative charge on the core Fe atoms.

The electrochemical behavior of the other cluster types is easily summarized by comparison to the results obtained for clusters **2**. Thus, as expected based on conclusion *(iii)* above, complexes **1** exhibit virtually identical redox properties to complexes **2** with the same terminal thiolate ligands. This is noteworthy in that the intracluster distance in **2** (ca. 3.67 Å) is significantly longer than the two distances reported for complexes of type **1** (3.19, 3.31 Å[76]), suggesting that the intercluster coupling is not strongly dependent on distance. Complexes of type **4** exhibit a single one-electron reduction at *ca.* $-0.9$ V, corresponding to reduction of the bridging ferric iron, and two overlapping, barely resolved one-electron reductions, corresponding to reduction of each of the $MoFe_3S_4$ cores. The potentials for the second two reductions are separated by $<100$ mV, in agreement with the large intracluster separation (6.6–6.9 Å[113]), and occur at significantly more negative potentials (by $\sim 200$ mV) than the first two reductions for clusters of type **2** with the same thiolate ligand (possibly due to electrostatic effects).

These results show that clusters **4** and **5** are related by one electron as the first two members of an at least four-membered electron transfer series:

$$\underset{\mathbf{4}}{[Mo_2Fe_7S_8(SR)_{12}]^{3-}} \overset{e^-}{\rightleftharpoons} \underset{\mathbf{5}}{[Mo_2Fe_7S_8(SR)_{12}]^{4-}} \overset{e^-}{\rightleftharpoons} [Mo_2Fe_7S_8(SR)_{12}]^{5-} \overset{e^-}{\rightleftharpoons} [Mo_2Fe_7S_8(SR)_{12}]^{6-} .$$

These systems containing two $MoFe_3S_4$ cores thus have the potential ability to act as two-electron reductants capable of delivering two electrons at very nearly the same potential. This capability has been utilized in the development of a photochemically driven hydrogen-evolving system, which uses $[Mo_2Fe_6S_8(SCH_2CH_2OH)_9]^{3-}$ in aqueous solution to shuttle electrons between illuminated spinach chloroplasts and clostridial hydrogenase[139]. In addition to their ability to act as two electron transfer agents, the reduced (net charge $5^-$) clusters **2** apparently can interact directly with reduceable substrates. Thus, simple treatment of pure $[Mo_2Fe_6S_8(SPh)_q]^{5-}$ in aprotic solvents with excess thiophenol results in evolution of hydrogen with concomitant oxidation of the cluster, as shown in the reaction below[138]:

$$[Mo_2Fe_6S_8(SPh)_q]^{5-} + 2\,PhSH \rightarrow [Mo_2Fe_6S_8(SPh)_q]^{3-} + 2\,PhS^- + H_2\uparrow .$$

In the presence of a large excess of thiophenol, the reaction is relatively rapid and almost quantitative. This is in marked contrast to results with two equivalents of $[Fe_4S_4SPh)_4]^{3-}$, which, although having the same redox potential and total reducing capacity, give only sluggish evolution of low yields (10–30%) of $H_2$ from excess thiophenol[138]. These results reinforce the idea that the reduced double cubanes can transfer two electrons intramolecularly to substrate, presumably generating a transient metal-hydride species that is protonated to give $H_2$.

*b. Ligand Substitution Reactions.* By analogy to the established reactivity of the $[Fe_4S_4(SR)_4]^{2-}$ clusters with thiols[140], phenols[141], acyl halides[36, 142] and other electrophiles[142], whch can be described by the following reaction

$$[Fe_4S_4(SR)_4]^{2-} + nEX \leftrightarrows [Fe_4S_4(SR)_{4-n}X_n]^{2-} + nRSE,$$

it might be expected that the $MoFe_3S_4$-containing clusters **1–5** would also undergo ligand exchange reactions of the terminal thiolates. In these systems, there clearly exists the possibility of selective reactivity of terminal and bridging thiolates as well. To date, only clusters of types **2** and **4** have been examined. Initial reports suggested that reaction of **2** (R = Et) with excess 2-hydroxyethanethiol resulted in complete substitution of both terminal and bridging thiolates to give **2** (R = $CH_2CH_2OH$)[114], while reaction with acyl halides resulted in selective substitution of terminal thiolates[108, 143] giving complexes of type **6**: $[Mo_2Fe_6(\mu_2\text{-}SR)_3X_6]^{3-}$ (X = Cl, Br). A more extensive investigation[144] has confirmed the latter reaction, but has shown that the former claim is apparently incorrect. Comparison of properties of independently synthesized $[Mo_2Fe_6S_8(SCH_2CH_2OH)_q]^{3-}$ with those of the product of the reaction of **2** (R = Et) with excess 2-hydroxyethanethiol and the use of $HSCD_2CH_2OH$ in the reaction have shown conclusively that the product of the ligand exchange reaction is the mixed thiolate system **7**: $[Mo_2Fe_6(\mu_2-SR)_3(SR')_6]^3$ (R = Et, R′ = $CH_2CH_2OH$). Analogously, **2** (R = Et) reacts with excess thiophenol to give **7** (R = Et, R′ = Ph), and with acetyl chloride to give **6** (R = Et, X = Cl). We have recently found that **2** (R = Et) reacts with excess phenol to give phenoxide-ligated complexes such as **8**: $[Mo_2Fe_6(\mu_2-SR)_3(OR')_6]^{3-}$ (R = Et, R′ = Ph)[145]. Thus, ligand exchange reactions of complexes of type **2** appear to be confined to the terminal thiolates, with the bridging thiolates non-labile under the conditions examined:

$$[Mo_2Fe_6(\mu_2-SR)_3(SR)_6]^{3-} \xrightarrow{\text{xsEX}} [Mo_2Fe_6(\mu_2-SR)_3X_6]^{3-} (X = SR', \text{halide, OR; E = H, RCO}).$$

Similar reactivity is observed for clusters of type **4**[145]:

$$[Mo_2Fe_7(\mu_2-SR)_6(SR)_6]^{3-} \rightarrow [Mo_2Fe_7(\mu_2-SR)_6X_6]^{3-} .$$

Use of large excesses of acyl halides results in fragmentation of **2** and **4**[144]. The observed inertness of the bridging thiolates is not suprising, as coordination to two metal centers would be expected to substantially reduce the nucleophilic character of the thiolate sulfur. Additionally, all reasonable assignments of oxidation states indicate that the Mo atoms in **1–5** are approximately trivalent; octahedrally coordinated Mo(III) ($d^3$) would be expected to be kinetically non-labile.

*c. Bridge Cleavage Reactions.* The weight of evidence indicates that FeMo-co contains only a single Mo atom per functional unit, and the Mo is thought to participate in catalysis. Because the existence of the triply bridged $Mo(SR)_3M$ unit in **2–8** provides a coordinatively saturated and substitution-inert pseudooctahedral coordination geometry at Mo, a great deal of interest centers on the generation of single $MoFe_3S_4$ clusters that may contain a relatively labile and reactive Mo site. Aside from an unsubstantiated report of the synthesis of $[(t\text{-}BuS)MoFe_3S_4(SPh)_3]^{2-}$[146], the only route to such species appears to be via cleavage of the double cubane species obtained by direct synthesis. Because of the apparent inertness of the $Mo(SR)_3Mo$ unit, attention has been focussed on the Fe-bridged species **4** and **5**, in the hope that the usual lability of iron thiolate ligation would allow cleavage at that point.

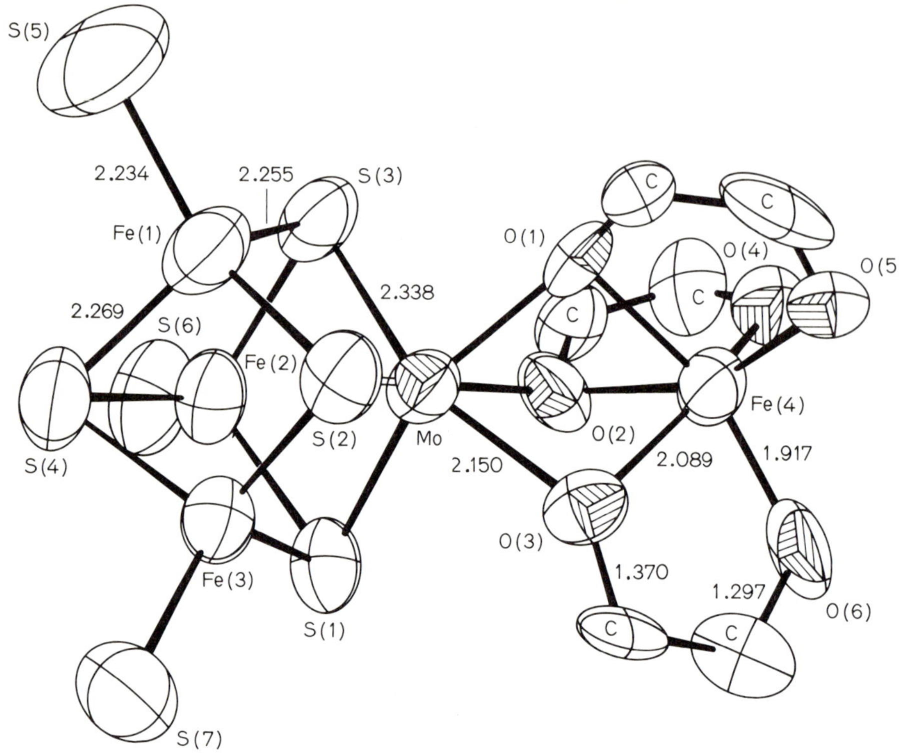


**Fig. 6.** The structure of the "catechol-capped" complex containing a single $MoFe_3S_4$ cubane core, **9**. The ethyl groups of the three thiolate ligands and four ring carbons of each of the three catecholate ligands have been omitted for clarity. Reproduced with permission from Ref.[116]; copyright by the American Chemical Society

Reaction of the Fe(III)-bridged species **4** (R = Et) with excess catechol (a ligand with a very high affinity for Fe(III)[147]) results in smooth cleavage of the double cubane to afford a new species containing a single $MoFe_3S_4$ core, $[MoFe_3S_4(SEt)_3Fe(C_6H_4O_2)_3]^{3-}$ (**9**)[116]. The structure of **9**, shown in Fig. 6, consists of an $MoFe_3S_4$ core to the Mo atom of which a tris(catecholate)iron(III) complex is attached via three bridging catecholate oxygens. Thus, although the goal of attaining a single $MoFe_3S_4$ unit has been attained, that of achieving coordinative unsaturation at Mo has not. The dimensions of the $MoFe_3S_4$ core are virtually identical with those of the corresponding units in **1–5**. In addition, the $^{57}Fe$ isomer shift in the Mössbauer spectrum (0.31 mm/s) and the isotropic shifts observed in the proton NMR spectrum ($SCH_2$- at −57 ppm from TMS at 22 °C) are virtually identical to the parameters obtained for complexes **1**, **2**, **4** and **5**. These results strongly suggest that the overall oxidation level of the $MoFe_3S_4$ core is unchanged, and that the average oxidation state of Fe remains unchanged at *ca.* +2.67[116]. Use of the more sterically hindered 3,5-di-*t*-butyl-catechol, which might be unable to accomodate a structure such as **9**, apparently gives a different product[116], the identity of which has not been reported. Also unknown is the fate of the other $MoFe_3S_4$ core of **4** upon treatment with catechol, and whether the $Fe(C_6H_4O_2)_3$ unit of **9** can be removed to allow chemistry to be carried out at the Mo site.

## 5. *Tungsten Analogs*

One of the goals of the research described above is to identify those structural and electronic features of FeMo-co that are essential for its activity. In particular, an especially puzzling aspect of the nitrogenase system to an inorganic chemist is the apparent inability of tungsten to replace molybdenum in functional nitrogenase[106, 148]. Accordingly, the tungsten analogs of virtually all $MoFe_3S_4$ compounds discussed above have been prepared and characterized, in the hope that some significant differences between the properties of Mo-Fe-S and W-Fe-S clusters will emerge. The properties of the latter will be summarized briefly and compared to those of the $MoFe_3S_4$ complexes discussed in more detail earlier.

The crystal structures of three tungsten double cubane clusters have been reported[112, 113, 117]. In all three cases (cluster types **2**[117], **3**[112], and **4**[113]), the tungsten analogs are isomorphous and virtually isostructural with the molybdenum complexes. In particular, the core dimensions show no chemically significant differences. The only consistent difference seems to be a very slight (0.01–0.03 Å) increase in the M–M distance of the triply bridged unit upon replacing Mo by W. The two main features of the electronic spectrum are shifted to higher energy by $\leq 10$ nm and 10–25 nm for the lower and higher energy bands, respectively, in the W analogs[108]. Isotropic shifts in the proton NMR spectra[108, 111] and $^{57}Fe$ isomer shifts in the Mössbauer spectra[108, 130] are virtually identical in the Mo- and W-containing species, while quadrupole splittings are typically 0.1–0.2 mm/s greater for the W clusters[108, 130]. The limited magnetic data available also suggest that the corresponding Mo and W systems are extremely similar to one another[134]. Finally, electrochemical studies[108, 117, 130] indicate that the $WFe_3S_4$ clusters **1–3** exhibit the same pattern of oxidation and reduction steps found for the $MoFe_3S_4$ clusters, with a negative shift of $\leq 100$ mV observed for all electrochemical processes in the W analogs. Clusters **4** and **5** seem to be only slightly more sensitive to W substitution, with a + 130 mW shift of the potential for reduction of the bridging Fe in **4** and a slight increase in the separation of the first two reductions of **5** being observed[108]. Only two significant differences have been observed to date, and both seem to reflect a higher affinity of W for oxygen donor atoms. The first is that all attempts at preparing the tungsten analogs of **2** with aryl thiolate ligands have apparently failed, yielding instead double cubane clusters **3** with three bridging methoxide ligands derived from the methanolic reaction medium[111, 112]. In contrast, synthesis of the methoxide-bridged double cubane **3**, M = Mo, required development of a separate synthetic procedure. The second is the observed lack of reversibility of electrochemical reduction of $WFe_3S_4$, but not $MoFe_3S_4$, clusters in oxygen donor solvents such as DMSO[130]. This has been attributed[115] to the sensitivity of the reduced $WFe_3S_4$ systems to attack by the oxygen-containing solvent (DMSO is a reasonably good two electron oxidant in some systems[149]). It has been suggested[115] that by analogy the putative W-Fe-S cluster in a tungsten analog of FeMo-co might be chemically less robust and therefore sensitive to attack by oxygen donors, and that this may be the origin of the observed lack of nitrogenase activity in organisms grown on tungstate in place of molybdate[106, 148].

## *C. Linear $FeS_2Mo$ Clusters*

### *1. Synthesis*

The ability of the tetrahedral tetrathiometallate ions, $[MS_4]^{2-}$ (M = Mo, W), to act as bidentate ligands toward a variety of divalent transition metal ions is well documented[93]. Thus, well-characterized bis(bidentate) complexes of the type $[M'(MS_4)_2]^{2-}$ have been prepared in which M′ = Ni[150–152], Zn[151, 152], Co[152], Pd[153], and Pt[153], while X-ray structure determinations for salts of $[Ni(MoS_4)_2]^{2-}$ [154], $[Co(WS_4)_2]^{2-}$ [155], and $[Zn(WS_4)_2]^{2-}$ [156] have been reported. Apparently analogous complexes $[Fe(MS_4)_2]^{2-}$, have only been obtained in impure form[151, 157]. These simple $[M'(MS_4)_2]^{2-}$ complexes are generally prepared in aqueous solution from the tetrathiometallate and an appropriate salt of M′ in a 2:1 mole ratio, followed by precipitation with a suitable bulky cation (e.g., $R_4N^+$, $Ph_4P^+$).

In contrast, reaction of the tetrathiometallates with excess (3–4 eq) $FeCl_3$ and thiolate in non-aqueous solution produces the double cubane clusters described above. The use of more reduced forms of iron and lower Fe: $[MS_4]^{2-}$ ratios (*ca.* 1–2 : 1) results instead in the formation of complexes containing $FeS_2Mo$ bridging units[158]. To date, five types of anionic clusters of this general type have been reported, including both binuclear and trinuclear species: $[S_2MoS_2FeX_2]^{2-}$, **10a**, X = SR[159, 160]; **10b**, X = Cl[160–163]; **10c**, X = OAr[164]; **10d**, X = NO[165]; **10e**, $X_2 = S_5$[166]; $[Cl_2FeS_2MoS_2FeCl_2]^{2-}$, **11**[167, 168]; $[S_2MoS_2FeS_2Fe(SAr)_2]^{3-}$, **12**[169, 170]; $[Fe(S_2MoS_2)_2]^{3-}$, **13**[171, 172]; $[(RS)_2MoS_3FeS_3Mo(SR)_2]^{3-}$, **14**[173]. Their structures are shown schematically in Fig. 7. Complexes **10–13** contain discrete bidentate tetrathiomolybdate moieties attached to tetrahedral Fe sites, while **14** contains square pyramidal Mo(V) with one terminal and two bridging sulfide ligands.

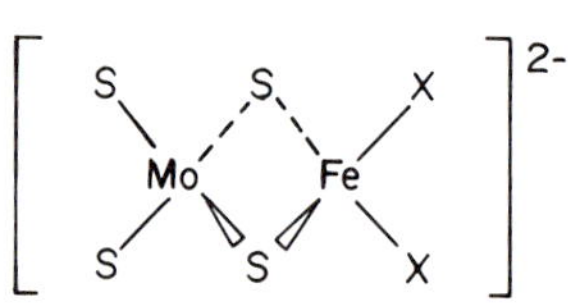

*10*

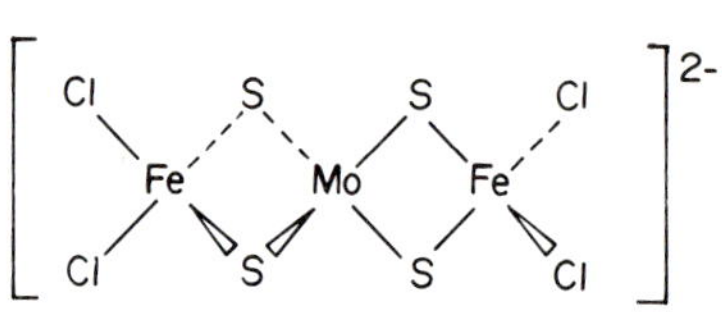

*11*

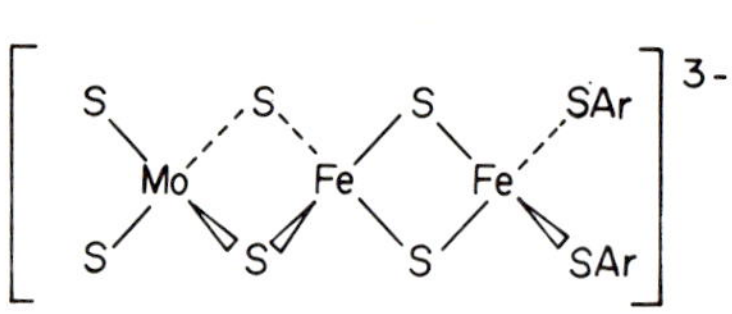

*12*

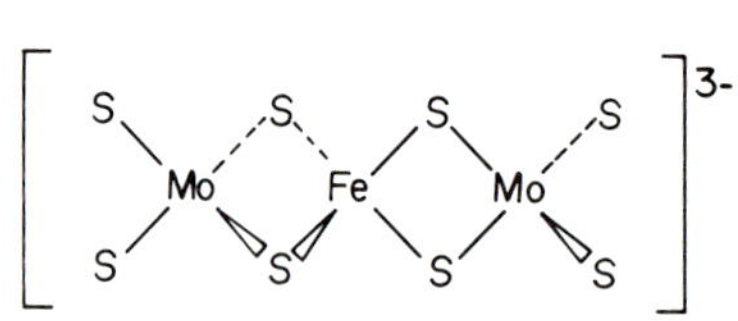

*13*

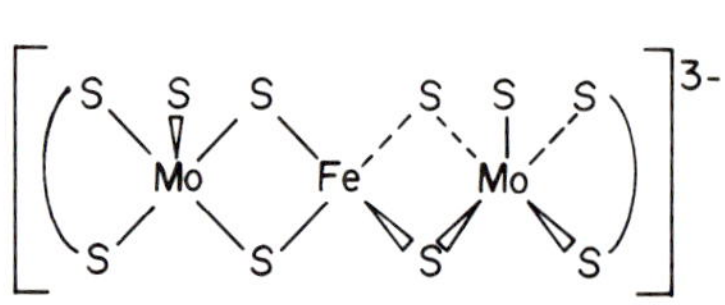

*14*

**Fig. 7.** Schematic of the structurally characterized linear $MoS_2Fe$ clusters prepared to date

As might be anticipated given the diversity of structures **10–14**, the situation that obtains with the double cubane clusters, namely that all or many of the known complexes can be isolated from a single reaction system[107)], does not obtain here. Instead, each complex **10–14** is the product of a specific set (or sets) of experimental conditions, and reported syntheses range from simple assembly reactions of mononuclear complexes to degradation reactions of higher nuclearity clusters. Thus, **10 b** and **11** are prepared simply by reaction of $[MoS_4]^{2-}$ with one[160, 161)] or two[167, 168)] equivalents of $FeCl_2$, respectively, while **10 a** can be prepared from **10 b**[160)] or from $[Fe(SPh)_4]^{2-}$[159)] by straightforward ligand displacement reactions. Complex **13** appears to be a very stable entity that forms spontaneously from $[MoS_4]^{2-}$ and a variety of Fe sources under reducing conditions[160, 171, 172)]. It is one electron more reduced than the typical $[M'(MS_4)_2]^{2-}$ clusters, and may be viewed as containing either Fe(I) and two Mo(VI) or Fe(III) and two Mo(V) as extreme assignments of oxidation states. The trinuclear cluster **12** can be obtained by reaction of $[MoS_4]^{2-}$ with either $[Fe_2S_2(SAr)_4]^{2-}$[169)] or $[Fe_4S_4(SAr)_4]^{2-}$[170)], while **14** is produced from an ill-defined reaction mixture containing $FeCl_3$, $[MoS_4]^{2-}$, and excess ethane-1,2-dithiolate[173)]. Other routes to these clusters, including the synthesis of **10 c–e**, will be discussed below under reactivity.

## 2. *Structures*

Single crystal X-ray diffraction structural determinations have now been performed for salts of some nine different linear cluster anions containing the $MoS_2Fe$ unit, including **10 a** (R = Ph)[159, 160)], **10 b**[160, 161, 163)], **10 c** (Ar = Ph)[174)], **10 d**[165)], **10 e**[166)], **11**[167)], **12** (Ar = *p*-Tol)[170)], **13**[172)], and **14**($(RS)_2 = SCH_2CH_2S$)[173)]. Typical values of structural parameters are summarized in Table 2 (see also[158)]), together with data for known species containing the $FeS_2Fe$ core.

Examination of Table 2 and the published data reveals substantial structural similarities among these clusters. Thus, for complexes with exclusively sulfur ligation to iron (**10 a**, **10 e**, **12–14**), the Mo-Fe distances vary within the relatively narrow range of 2.73–2.76 Å. Coordination of more electronegative groups (Cl, OAr, $NO^+$) to iron (**10 b–d**, **11**) results in significant lengthening of the Mo-Fe distance (to 2.835 Å in **10 d**[165)]). Although these distances are within the range expected for metal-metal covalent single bonds, the fact that very similar distances are observed in structurally characterized $[M'(MS_4)_2]^{2-}$ complexes for M' = Zn, Ni, Co[154–156)], in which direct M-M' bonding is very unlikely, suggests that they may simply be a result of geometrical constraints of the tetrathiometallate ligand[158)]. The Mo-S* (bridging) and Mo-S (terminal) distances likewise fall into narrow ranges, $2.25 \pm 0.01$ Å and $2.16 \pm 0.02$ Å, with average values only very slightly longer than for $[MoS_4]^{2-}$ itself (2.18 Å)[121, 122)], suggesting a valence near + 6 for Mo. Exceptions are **11**, in which all four S atoms of $[MoS_4]^{2-}$ are bridging (average Mo-S 2.204 Å[167)]) and **14**, which has been formulated as containing Mo(V), consistent with the long Mo-S* distance of 2.322 Å[173)]. Although the Mo-Fe distances observed for the linear clusters are comparable to those found by EXAFS for the Mo site of nitrogenase (2.72 Å[69–71)]), the Mo-S distances observed in the linear clusters are substantially shorter than the 2.35 Å found by EXAFS for the four S nearest neighbors of Mo in nitrogenase. This reinforces the idea of a relatively reduced Mo site in the FeMo-cofactor, and suggests that more accurate linear cluster models will have to utilize Mo in oxidation states less than + 6 or + 5.

**Table 2.** Selected average structural parameters for linear $MoS_2^*Fe$ clusters and related units[a]

| | **10a**[b] | **11**[c] | **12**[d] | **13**[e] | **14**[f] | **10a**-W[g] | $[Fe_2S_2(S\text{-}p\text{-}Tol)_4]^{2-}$ [h] | $[Fe_2S_2Cl_4]^{2-}$ [i] |
|---|---|---|---|---|---|---|---|---|
| Mo-Fe | 2.756(1) | 2.775(6) | 2.778 | 2.740(1) | 2.746(5) | 2.772(1) | | |
| Fe-Fe | | 5.551(2) | 2.691 | | | | 2.691(1) | 2.716(1) |
| Mo-S* | 2.255(2) | 2.204(5) | 2.244 | 2.256(5) | 2.322(9) | 2.245(4) | | |
| Fe-S* | 2.264(2) | 2.295(5) | 2.278<br>2.212[j] | 2.256(11) | 2.221(22) | 2.287(2) | 2.201(1) | 2.201(1) |
| Mo-S | 2.153(2) | [k] | 2.166 | 2.171(3) | 2.139(9)[l] | 2.159(2) | | |
| Fe-X | 2.307(2) | 2.225(10) | 2.321 | | | 2.305(3) | 2.312(1) | 2.252(1) |
| S*---S*′ | 3.576(2) | 3.543(6) | [m] | 3.579(5) | [m] | 3.581(2) | 3.483(3) | 3.463 |
| Mo-S*-Fe | 75.18(6) | 76.05(1) | 75.81<br>74.92[n] | 74.77(13) | [m] | 75.41(16) | 75.39(4)[n] | 76.21(3)[n] |
| S*-Mo-S* | 104.97(6) | 109.5(2) | 105.29 | 105.0(2) | 98.5(1.6) | 105.87(6) | | |
| S*-Mo-S′ | 111.34(8) | | 110.99 | [m] | | 110.58(9) | | |
| S*-Fe-S*′ | 104.33(7) | 100.9(2) | 103.35<br>106.36 | 104.9(2) | 104.7(2.2) | 103.03(6) | 104.61(4) | 103.79(3) |
| X-Fe-X′ | 105.44(7) | 110.2(9) | 109.19 | | | 106.35(8) | 111.2 | 105.37(4) |
| M-M′-M″ | | 179.38(7) | 176.37 | 172.64(6) | 155.7(2) | | | |

[a] Distances in Å, angles in deg. S* denotes bridging sulfide.
[b] Ref.[160].
[c] Ref.[167].
[d] Ref.[170].
[e] Ref.[172].
[f] Ref.[173].
[g] Ref.[174, 177]; distances and angles for W rather than Mo.
[h] Ref.[175].
[i] Ref.[176].
[j] Values for central, terminal.
[k] All Mo-S are bridging.
[l] Terminal Mo-S; also 2 terminal Mo-S (mercaptide) at 2.402(9) Å.
[m] Not reported.
[n] Fe-S*-Fe′ angle

The overall dimensions and stereochemistry of the $S_2MoS_2FeX_2$ units in **10** are very similar to those of the corresponding $[Fe_2S_2X_4]^{2-}$ (X = S-*p*-Tol, Cl) clusters[175, 176], and a convenient view of the former is to consider them as arising from substitution of one of the Fe atoms of the $Fe_2S_2$ unit by Mo (much as the $MoFe_3S_4$ cubane can be viewed as arising from substitution of one Fe atom of an $Fe_4S_4$ core by Mo). Thus, the acute angles at the bridging S atoms, the bridging S---S distances, terminal ligand-Fe distances, and the approximately tetrahedral coordination at each metal are essentially identical in the two systems. Incorporation of Mo results in a slight ($\sim 0.05$ Å) increase in the metal-metal distance (presumably due to the larger covalent radius of Mo vs. Fe) and an increase of *ca.* 0.05 Å in the Fe-S (bridging) distance. Within the series of complexes **10–14**, the sum of the Mo-S (bridging) and Fe-S (bridging distances is virtually constant at $5.52 \pm 0.02$ Å.

Of the trinuclear clusters **11–14**, complexes **11** and **12** illustrate most clearly the potential for elaboration of the basic $MoS_2Fe$ unit. Thus, **11** contains $[MoS_4]^{2-}$ coordinated to two $FeCl_2$ moieties, and provides the closest synthetic approach yet to two of the structural models proposed for FeMo-co (Fig. 2, structures **III** and **IV**), although the short Mo-S distances in **11** rule it out as a detailed model for the $FeS_2MoS_2Fe$ portion of the cofactor. Nonetheless, **11** is the only complex prepared to date with two Fe atoms as nearest neighbors to Mo, and comparison of its Mo EXAFS spectrum (not yet reported) to those of the $MoFe_3S_4$ clusters **1** and **2** may resolve the question of the presence of 2 or 3 Fe atoms as neighbors to Mo in nitrogenase. Complex **12** demonstrates that one of the key features of structural model **III** (Fig. 2), namely, coordination of an $S_2MoS_2$ unit to an example of a familar $Fe_nS_n$ core, is chemically feasible. In this particular case, coordination of $[MoS_4]^{2-}$ has stabilized a 2 Fe-2 S core in the reduced (Fe(II)-Fe(III)) mixed valence state (see below) previously unknown in synthetic complexes.

### 3. *Properties*

In polynuclear complexes such as **10–14**, a key question concerns the distribution of valence electrons among the metal ions (assignment of metal oxidation states). For even the simple binuclear units **10 a–c**, there are two possible extremes: Fe(II)-Mo(VI) or Fe(III)-Mo(V). The trinuclear systems **11–14** are obviously more complicated yet, but offer the same range of possible oxidation sttes in most cases. In addition, **13** can be formulated as containing Fe(I), and **10 d** as containing Fe(0) as one extreme! In order to examine this problem more closely, a variety of spectroscopic and other physical probes have been employed; pertinent results from these studies will be summarized below. It is worth noting again that, with the exception of **14**, the structural results discussed above are consistent with assignment of Mo as Mo(VI) in all these compounds, with no evidence for substantial change transfer to or reduction of Mo.

*a. Electronic Spectra.* In contrast to the relatively undistinguished optical spectra of the $MoFe_3S_4$ cubane clusters, the visible spectra of **10–14** exhibit a number of relatively narrow, intense ($\varepsilon \sim 10{,}000$–$20{,}000$ $M^{-1}$ $cm^{-1}$) bands (Fig. 8); in addition, broad, lower intensity ($\varepsilon \sim 100$ $M^{-1}$ $cm^{-1}$) bands have been observed in the near-infrared region for several examples of **10**[158, 160, 177]. The best studied of these systems are **10 a, b**[158, 160], the visible spectra of which are dominated by a pair of intense bands between 400 and

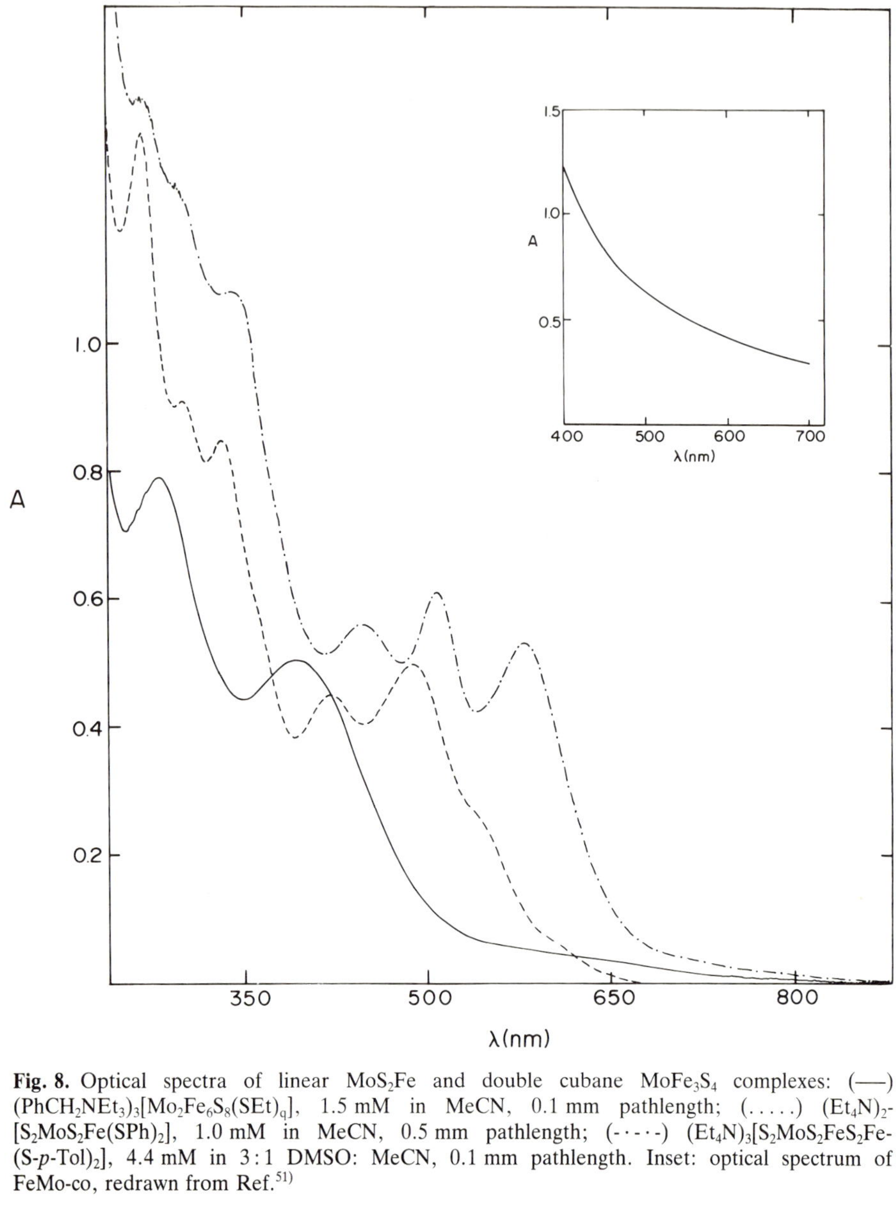


**Fig. 8.** Optical spectra of linear $MoS_2Fe$ and double cubane $MoFe_3S_4$ complexes: (—) $(PhCH_2NEt_3)_3[Mo_2Fe_6S_8(SEt)_q]$, 1.5 mM in MeCN, 0.1 mm pathlength; (.....) $(Et_4N)_2$-$[S_2MoS_2Fe(SPh)_2]$, 1.0 mM in MeCN, 0.5 mm pathlength; (-····-) $(Et_4N)_3[S_2MoS_2FeS_2Fe$-$(S$-$p$-$Tol)_2]$, 4.4 mM in 3:1 DMSO: MeCN, 0.1 mm pathlength. Inset: optical spectrum of FeMo-co, redrawn from Ref.[51)]

500 nm. These features are similar to those reported for complexes of the type $[M'(MoS_4)_2]^{2-}$[93, 152)], where coordination of $[MoS_4]^{2-}$ results in a lowering of the symmetry from $T_d$ and a roughly symmetrical splitting of the lowest energy (470 nm) S → Mo charge transfer band of $[MoS_4]^{2-}$. Weaker features are observed at both higher and lower energies; they may be due to iron-centered transitions, but a detailed assignment is not yet possible. The broad, weak bands between 900 and 1100 nm have been assigned as intramolecular mixed-valence charge transfer transitions, Fe(II) → Mo

(VI)[158, 177]; their observed half widths[177] are within 20% of those calculated according to classical theory describing such transitions[178]. Taken together, the electronic spectra strongly suggest that complexes **10 a–c** can be viewed as containing a tetrathiomolybdate (VI) perturbed by coordination to a reducing metal ion, Fe(II).

The presence of similar intense features in the visible spectra of **10 d, e, 11** and **12** can therefore be interpreted as indicating that these species also contain coordinated tetrathiomolybdate (VI). While the optical spectrum of **13** contains sharp bands of comparable intensity to those discussed above[171, 172], it is clearly more complex, with a distinct shift to longer wavelength of the first two intense features making simple interpretation impossible. Finally, complex **14** differs from the other linear clusters in exhibiting only relatively weak ($\varepsilon < 3000\ M^{-1}\ cm^{-1}$) bands in the visible spectrum, consistent with the absence of discrete $[MoS_4]^{2-}$ units, and its formulation as a Mo(V) complex[173].

*b. Vibrational Spectra.* All complexes **10–13** for which infrared spectra have been reported exhibit two bands or groups of bands attributable to Mo-S stretching vibrations, at 480–505 and 430–460 $cm^{-1}$. This behavior is typical of $[MoS_4]^{2-}$ acting as a bidentate ligand[179], which results in a splitting of the parent $[MoS_4]^{2-}$ mode at 475 $cm^{-1}$ into components at higher and lower energies; these can be assigned to the Mo-S (terminal) and Mo-S (bridging) vibrations, respectively[179]. Resonance Raman spectra for **10 b**[162] and **11**[168] have been reported and are quite distinctive. It has been suggested that the resonance Raman effect may be of general use in the identification of the mode of coordination of tetrathiometallates in these and related systems[180].

*c. Proton Magnetic Resonance Spectra.* In contrast to the $MoFe_3S_4$ cubane clusters, proton magnetic resonance spectra of the linear $MoS_2Fe$ clusters have generally not been examined in great detail for two reasons: many of the complexes (e.g., **10 b, d, e, 11, 13**) have no organic constituent with observable protons, and the optical and infrared spectra are highly characteristic and thus useful for compound identification. The most detailed study is of the binuclear Mo-Fe complexes **10 a** with aryl thiolate ligands[160]. Large isotropic shifts (30–60 ppm) are observed for the aromatic protons; the pattern and relative magnitudes of the shift are consistant with dominant contact intractions. The temperature dependence of the isotropic shifts parallels that of the magnetic susceptibility (see below), consistent with simple Curie law behavior[159, 160].

*d. Mössbauer Spectra.* In view of the possible ambiguity discussed above in assigning oxidation states to the metal ions in these systems, $^{57}Fe$ Mössbauer spectra provide a direct means of investigating the electronic environment about the Fe atoms. Isomer shifts and quadrupole splittings for examples of complexes **10–13** are given in Table 3. Previous results[76, 131] have demonstrated a linear correlation between the isomer shift $\delta$ and the oxidation state of Fe in a tetrahedral sulfur environment, with values ranging from 0.30 mm/s for $Fe^{3+}$ to $\sim 0.70$ mm/s for $Fe^{2+}$. The isomer shift values for **10 a**, **10 e**, **13**, and **12** (doublet I) are thus consistent with an effective oxidation state of Fe of approximately +2.5, suggesting appreciable Fe → Mo charge transfer. Doublet II of **12** has parameters (both $\delta$ and $\Delta E_Q$) typical of high spin Fe(III) in a tetrahedral sulfur environment, and may be assigned to the Fe site more distant from the Mo in **12** (Fig. 7). Complexes with more electronegative ligands on Fe (e.g., **10 b, c, 11**) exhibit larger

**Table 3.** Mössbauer parameters at 4.2 K and room temperature magnetic moments of linear $MoS_2Fe$ clusters

| Complex | Isomer shift $\delta$ (mm/s)[a] | Quadrupole splitting $\Delta E_Q$ (mm/s) | Magnetic moment $\mu$ ($\mu_B$) | Ref. |
|---|---|---|---|---|
| **10a** (R = Ph) | 0.47 | 1.63 | 4.9, 5.1 | 158, 160 |
| **10b** | 0.60 | 2.12 | 5.3, 5.0 | 160, 162 |
| **10c** (Ar = Ph) | 0.57 | 1.20 | 4.9 | 164 |
| **10e** | 0.51[b] | 1.66[b] | 4.9 | 166 |
| **11** | 0.58 | 1.99 | 6.6 | 167 |
| **12** (Ar = Ph) | I 0.42<br>II 0.30 | 1.41<br>0.68 | 1.7 | 169 |
| **13** | 0.42[b] | 1.04[b] | 3.9 | 158, 172 |
| **14** | | | 4.4 | 173 |
| **10d** | | | 0 | 158 |

[a] Relative to metallic Fe at room temperature
[b] At 77 K

isomer shifts and quadrupole splittings, indicating significantly greater ferrous character and therefore less electron donation from Fe to Mo. It should be noted, however, that a detailed analysis of the Mössbauer spectra of **10a** (R = Ph) in applied longitudinal magnetic fields of up to 60 kOe has shown that the results may be interpreted in terms of high-spin $Fe^{2+}$ [160], suggesting that a simple relationship of oxidation state to isomer shifts in these $[MoS_4]^{2-}$-based complexes may not exist.

*e. Magnetic Susceptibility.* The magnetic data presented in Table 3 demonstrate that the linear $MoS_2Fe$ clusters exhibit a variety of spin states. Thus, the room temperature magnetic moments reported for the dimeric complexes **10a–c** and **10e** are very near the spin-only value of 4.93 $\mu_B$ per formula unit, consistent with a high-spin Fe(II)-Mo(VI) formulation[161]. It is impossible, however, to distinguish by magnetic measurements between this and an alternative involving substantial charge transfer from Fe to Mo, if the two metals are strongly antiferromagnetically coupled. The magnetic susceptibility of **10a** (R = Ph) has been examined from 4.2 to 300 K[160], and the data are completely consistent with simple Curie law behavior for an S = 2 system above 10 K, with a weak *inter*molecular antiferromagnetic interaction observed in the solid state below 10 K. The magnitude and temperature dependence of the magnetic moment of the closely related trinuclear cluster **11** are consistent with a weak antiferromagnetic interaction of two high-spin Fe(II) ions via the bridging $S_2MoS_2$ unit[158, 167].

The other trinuclear clusters **12–14** are apparently characterized by extensive intramolecular antiferromagnetic coupling as well. Thus, the reported magnetic moment of 1.7 $\mu_B$ for the $[S_2MoS_2FeS_2Fe(SPh)_2]^{3-}$ ion (**12**), together with its Mössbauer spectrum, suggests that it is best formulated as containing high-spin Fe(II) (S = 2) and high-spin Fe(III) S = 5/2), antiferromagnetically coupled to give an S = 1/2 ground state[169, 170]. The magnetic properties of both **13** and **14** are consistent with an S = 3/2 ground state[158, 172, 173], which could arise either from tetrahedral high-spin Fe(I) with two Mo(VI) atoms or from high-spin Fe(III) antiferromagnetically coupled to two S = 1/2 Mo(V) atoms. The latter formulation seems to be in substantially better agreement with the Mössbauer parameters for **13** and the structural data for **14**[173]. Finally, the $[S_2MoS_2Fe(NO)_2]^{2-}$ ion is diamagnetic, and has been viewed as formally containing $NO^+$ and $d^{10}$ Fe(2–)[158].

*f. Electron Paramagnetic Resonance Spectra.* Only two of these complexes exhibit well-resolved EPR spectra. A narrow, isotropic signal observed at g = 2.005 for the trinuclear complex **12** at low temperatures is consistent with an S = 1/2 ground state[169], but a detailed description of the electronic properties of the complex remains to be developed. The $[Fe(MoS_4)_2]^{3-}$ ion shows a rhombic S = 3/2 EPR spectrum that is very solvent dependent and, under certain conditions, is somewhat similar in apperance to that of FeMo-co[171]. For example, in frozen aqueous solution, the apparent g values are 5.3, 2.6, and 1.7[181]. If complex **14** also proves to have an S = 3/2 ground state, a somewhat similar EPR spectrum at low temperature would be expected as well.

## 4. *Reactivity*

*a. Oxidation-Reduction Reactions.* The ability of the tetrathiometallate ligands to stabilize relatively reduced oxidation states of the central metal ions M′ in complexes such as $[M'(MS_4)_2]^{n-}$ is demonstrated by the existence of the three-membered electron transfer series

$$[M'(MS_4)_2]^{2-} \leftrightarrows [M'(MS_4)_2]^{3-} \leftrightarrows [M'(MS_4)_2]^{4-} ,$$

where M′ = Ni, Pd, Pt; M = Mo, W[153] and M′ = Co, M = W[182], and by the isolation of $[Fe(MoS_4)_2]^{3-}$[160, 171, 172] rather than the corresponding dianion even under only mildly reducing conditions. Consequently, a relatively rich electrochemistry might be anticipated for the linear $MoS_2Fe$ clusters **10–14**; unfortunately, this has not proven to be the case. The binuclear systems **10a–c** exhibit only irreversible reductions at about – 1.3 to – 1.5 V vs. the standard calomel electrode (SCE), with diffusion currents corresponding roughly to a one electron reduction. Attempts at chemical generation of a reduced species with 1–3 equivalents of the acenaphthalene radical anion ($E_{1/2}$ = – 1.62 V vs SCE) result instead in virtually quantitative (based on Mo) formation of $[Fe(MoS_4)_2]^{3-}$[177], suggesting that the reduced forms of **10a–c** are unstable, rearranging rapidly to **13**. A reversible one electron reduction has been claimed for the trinuclear system **11**[158], but no further electrochemical data are given. Complex **12**, which contains a half-reduced $Fe_2S_2$ core ligated by one $[MoS_4]^{2-}$, exhibits only an irreversible oxidation at + 0.05 V and an irreversible reduction at ca. – 1.1 V vs. SCE[169, 170]; presumably, the reduced species undergoes a chemical rearrangement analogous to **10a**. Quite surprisingly, complex **13**, $[Fe(MoS_4)_2]^{3-}$, shows only a quasireversible one electron reduction at very negative potentials (ca. – 1.75 V vs SCE)[171, 172]; a one electron oxidation at – 0.075 to – 0.10 V vs SCE is electrochemically irreversible[171, 172]. Since this oxidation should produce $[Fe(MoS_4)_2]^{2-}$, which is analogous to other stable bis(tetrathiomolybdate) complexes, this result is unexpected. Studies on the tungsten analog, however, have shown that the dianion adds two oxygen donor solvent molecules readily, and similar behavior may account both for the irreversibility of the oxidation of **13** and for the difficulty in preparing pure $[Fe(MoS_4)_2]^{2-}$ in the past[151, 157]. Complex **14** is reported to undergo a quasireversible one electron reduction at relatively mild potentials (– 0.67 V vs SCE)[173]; this may be attributed to reduction of the central Fe(III). The available data thus suggest that reduction of these complexes does not, in general, generate stable reduced species; instead, relatively reactive species result that undergo further reaction with solvent or with themselves.

*b. Ligand Substitution Reactions.* Only for the binuclear complexes **10** has the reactivity of the terminal ligands on Fe been examined in any detail[158, 160]. Results obtained to date are summarized in Fig. 9, and indicate that the reactivity of the Fe portion of the $MoS_2Fe$ unit parallels that of the $Fe_2S_2$ unit[142]. Thus, reaction of the arylthiolate complexes **10a** with acyl chlorides results in formation of the chloro analog **10b**; the reaction can be reversed with arylthiolate. Similarly, reaction of **10b** with sodium phenoxide produces the phenoxide complexes **10c**[164]; treatment of **10c** with arylthiol or acyl chloride yields **10a** or **10b** respectively. An important point is that these ligand substitution reactions are virtually quantitative[160], indicating that the $MoS_2Fe$ core is inert under these conditions. The $S_5^{2-}$ ligand is introduced by a similar route, involving treatment of **10a** with benzyltrisulfide[166] or elemental sulfur[177].

The reactivity of the trinuclear clusters **11–14** is virtually unexplored. It has been shown[167] that $[Cl_2FeS_2MoS_2FeCl_2]^{2-}$ dissociates in solvents such as DMF to $FeCl_2$ and $[S_2MoS_2FeCl_2]^{2-}$, demonstrating that coordinated tetrathiomolybdate has only a low affinity for a second metal ion. Reaction of gaseous NO with $[Fe(MoS_2)_2]^{3-}$ does not give the expected NO adduct at the central iron, but instead results in cleavage of one $[MoS_4]^{2-}$ to give $[S_2MoS_2Fe(NO)_2]^{2-}$, **10d**[165].

*c. Other Reactions.* Because of the paucity of suitable soluble sources of Mo and S, the linear clusters, particularly **10a**, **b** and **12** which contain only a single Mo, would seem to be an obvious choice as starting materials for elaboration of more complex Mo-Fe-S units. To date, despite a great deal of effort (mostly unpublished) in several laboratories, these complexes have not provided a route to novel Mo-Fe-S clusters; it has only been possible to convert them into other, known cluster types. Thus, reaction of **10a** (R = Ph) with $FeCl_3$ and $PhS^-$ in methanol produces the double cubane cluster **2** (R = Ph) in excellent yield[115, 177]. Attempts to induce rearrangements of the $MoS_2Fe$ core by reaction of coordinated tetrathiomolybdate with electrophiles (e.g. MeI) or nucleophiles (e.g. MeLi) result in the conversion of **10b** to **11** and **13**, respectively, in good yield[177].

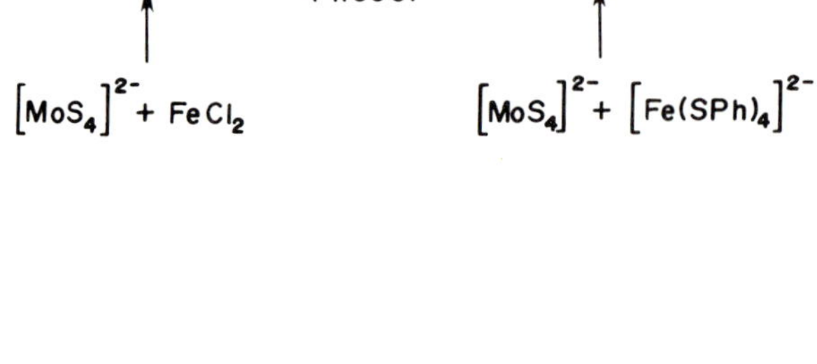


**Fig. 9.** Reactions of binuclear $FeS_2Mo$ clusters

### 5. *Tungsten Analogs*

As discussed above, comparative properties of W-Fe-S and corresponding Mo-Fe-S clusters are of appreciable interest, in view of the apparent inability of tungsten to replace molybdenum in a functional nitrogenase system. The tungsten analogs of complexes **10a**[177]), **10b**[158, 161, 162]), **10c**[174, 177]), **10d**[165]), **10e**[166]) **11**[167]), **12**[170]) and **13**[183]) have been prepared, and the crystal structures of three tungsten analogs of the linear $MoS_2Fe$ cluster anions (**10a** (R = Ph)[174]), **10b**[158, 161,] **10e**[166])) have been determined. In general, the trends in structural parameters and physical properties upon replacement parallel those observed for the $MFe_3S_4$ cubane clusters. Thus, the M-Fe distances increase only slightly, by ca. 0.015–0.03 Å, with the greatest increase observed for the chloride derivative **10b**. The intense tetrathiometallate-based transitions that dominate the optical spectrum are significantly blue shifted, as expected based on the spectra of the parent ions[93]) ($\lambda_{max}$ = 470, 396 nm for $[MoS_4]^{2-}$, $[WS_4]^{2-}$, respectively). Similar shifts are observed for other spectral features, including the near-infrared bands assigned to Fe → M excitations[158, 177]). Isotropic shifts in the proton NMR spectra of **10a, c**, and **12** are virtually unchanged in the W analogs, as are the magnetic properties of **10-W–13-W**. Mössbauer spectra of the tungsten analogs of **10a–c**, **10e**, and **11** show a small but significant increase (~ 0.03–0.06 mm/s) in isomer shift, consistent with less Fe → M charge transfer in the W complexes, and an increase of ~ 0.2 mm/s in the quadrupole splitting. No EPR data have been reported for $[Fe(WS_4)_2]^{3-}$, but the EPR spectra of W and Mo complexes **12** are virtually identical[170]). Electrochemical studies of **10** and **12** show only a negative shift of 50–100 mV for the irreversible processes discussed earlier. In contrast, $[Fe(WS_4)_2]^{3-}$ shows a *reversible* one-electron oxidation at − 150 mV vs. SCE[183]), suggesting that in this case the dianion expected by analogy to other bis(tetrathiometallates) is indeed stable. This is confirmed by its successful preparation and crystallization[183]). A structure determination, however, has shown that the dianion does not have the expected tetrahedral structure, but instead exists as the bis(dimethylformamide) adduct in which two DMF molecules are coordinated in the *trans* positions of a roughly octahedral Fe(II) ion[183]). This is accompanied by an increase in the Fe-W and Fe-S distances to 3.044 and 2.511 Å, respectively, suggesting a significant weakening of the Fe-$[MS_4]^{2-}$ interaction. This behavior is consistent with the higher affinity of W than Mo for oxygen ligands noted earlier.

## D. *String Bag Clusters*

Efforts directed at the synthesis of clusters related to the string bag model for FeMo-co (Fig. 2, **V**) or the *nido*-cubane $MoFe_3S_3$ fragment have utilized a variety of reductants rather than thiolates, including Grignard reagents[184, 185]) and a combination of metal hydrides plus ethyleneglycolate[186]), as well as *t*-butylmercaptide[187]), in reactions with $[MoS_4]^{2-}$, $FeCl_3$ or $FeCl_2$, and NaHS. The strategy is intended to eliminate or minimize the presence of bridging groups in the cluster products. Each of these systems appears to yield a variety of products. Thus, the series "G" complexes, isolated from reaction with Grignard reagents by recrystallization from DMF, all have the same unit cell parameters, and appear to be clathrate structures containing $[Mg(DMF)_6]^{2+}$ cations and $[S_2MoS_2FeCl_2]^{2-}$ or other related anions[184, 187]). The series "F" complexes, isolated as their

$Et_4N^+$ salts from reaction mixtures containing *t*-butylmercaptide, appear to be isomorphous with $(Et_4N)_2$ $[Fe_4S_4(S\text{-}t\text{-}Bu)_4]$; the Mo content has been attributed to a statistical distribution of Mo at the vertices of the cube[187]. Unfortunately to date no structural or spectroscopic data are available on these compounds, as the structures have been assigned based on elemental analyses and preliminary X-ray measurements only, and all of these complexes appear to be mixtures of several variants of one or more cluster types. In spite of their current structural ambiguity, these reaction systems are nonetheless of considerable interest, inasmuch as the products are all reported to exhibit some ability to reconstitute the defective MoFe protein from *A. vinelandii* mutant UW 45 with restoration of $C_2H_2$- and $N_2$-reducing capacity[184–187]. This is true of neither the double cubane $MoFe_3S_4$ nor the linear $MoS_2Fe$ clusters presently known, and suggests that a possibly minor constituent of the above reaction products may indeed approximate the structure of FeMo-co.

## E. Summary

Reaction of tetrathiomolybdate with either iron halide/mercaptide mixtures or preformed Fe-S clusters has resulted in the preparation of two basic types of synthetic Mo-Fe-S cluster: those containing the $MoFe_3S_4$ cubane core, and the "linear" clusters containing $MoS_2Fe$ units. Comparison of available structural data on representatives of each cluster type with EXAFS results on the Mo site of nitrogenase (the FeMo-cofactor) suggests that the $MoFe_3S_4$ cubane clusters, in which the Mo has a formal valence state of +3 to +4, more closely reproduce the Mo-Fe and Mo-S distances observed for FeMo-co than do the linear clusters, which contain Mo in a higher valence state (+5 to +6) and exhibit significantly shorter Mo-S bonds. Due to their coordinative unsaturation and the greater ease in constructing, at least conceptually, reasonable structures for FeMo-co containing 6–8 Fe atoms, the linear clusters would appear to have greater potential for generating FeMo-co models, *if reduction of the central Mo to a lower oxidation state can be achieved.* Despite extensive efforts in several laboratories, this has not been accomplished to date, and suggests that synthetic routes using sources of Mo other than $[MoS_4]^{2-}$ may be necessary. If, as current evidence suggests, there are indeed only 4 S atoms per Mo in FeMo-co and the Fe does indeed have low Z (O or N) atoms as ligands, then the current emphasis on Mo-Fe-S clusters will have to be adjusted to include non-sulfur (presumably, but not necessarily, oxide or hydroxide) ligands both within and external to the cluster core. To date, the only Mo-Fe-S complexes with exogenous oxygen ligands have employed phenoxide ligation to Fe, which appears to stabilize higher oxidation states compared to the sulfur analogs. Comparative data on the properties of Mo-Fe-S clusters with other types of oxygen ligands are not yet available. Finally, the effects observed upon replacing Mo by W are limited to very slight negative shifts in reduction potentials and to a somewhat higher affinity of tungsten for oxygen ligands; there would thus seem to be no obvious reasons for the instability or inactivity of an iron-tungsten cofactor.

# IV. Synthetic Fe-S Clusters as Models for the P Clusters

If one accepts the basic premise that the P clusters consist of tetranuclear spin-coupled iron-sulfur clusters in which one Fe is somehow differentiated from the other three in both the native ($P^N$) and one-electron oxidized ($P^{OX}$) states, then the question of whether the splitting of the Fe atoms into two spectroscopic environments in a 3:1 ratio is an inherent property of the $Fe_4S_4$ unit or is imposed upon it by the protein must be considered. A variety of evidence presented above in Sect. II.C.2 suggests that the diamagnetic $P^N$ state corresponds to a fully reduced (all ferrous) $[Fe_4S_4]^O$ unit, while the paramagnetic, $S \geq 5/2$, $P^{OX}$ state corresponds to some variant of the $[Fe_4S_4]^+$ unit, familiar from its occurrence in reduced 4Fe-4S ferredoxins, super-reduced high potential iron proteins, and the synthetic $[Fe_4S_4(SR)_4]^{3-}$ ions. Since the area of overlap between the P clusters and known Fe-S chemistry is at the $[Fe_4S_4]^+$ level, the relevant properties of the latter will be summarized briefly in order to assess the likelihood that the properties of the $P^{OX}$ state are those of an unconstrained $[Fe_4S_4]^+$ cluster with thiolate ligation. Evidence bearing on the properties of $Fe_4S_4$-clusters coordinated by physiologically relevant non-thiolate ligands will then be presented, and the relative merits of the two alternative explanations will be considered.

## A. Properties of $[Fe_4S_4(SR)_4]^{3-}$

The tetranuclear clusters $[Fe_4S_4(SR)_4]^{2-}$ exhibit two well-defined one-electron reduction processes, corresponding to reduction to the corresponding tri- and tetraanions; of these, the 2-/3-couple most closely approaches electrochemical reversibility[3,4,124]. For simple aliphatic and aryl thiolates, the potentials for the 2-/3- and 3-/4-couple range from − 1.0 to − 1.4 and − 1.75 to − 2.15 V vs. SCE[4,124]. A separation of ~0.75 V between 2-/3- and 3-/4-couples is observed for all tetranuclear clusters with thiolate ligands, while the separation increases slightly (to ~0.9 V) upon replacement of thiolate by halide ligands[142], suggesting that this constitutes an integral property of the $Fe_4S_4$ core. Even after correction to values versus the standard hydrogen electrode (add + 0.24 V), these potentials would appear to be sufficiently negative to preclude their existence in biological materials. A detailed study[125] has shown, however, that the differences in reduction potentials between synthetic and protein-bound $Fe_4S_4$ cores are minimal ($\leq$0.03 V) when both are measured by the same technique in mixed aqueous-organic solvents. These results would suggest that, given the normal range of − 300 to − 500 mV vs. SHE for the 2-/3-couple in 4Fe–4S proteins, the 3-/4-couple should have a potential of *ca.* − 1.0 to 1.3 V vs. SHE! The measured potential for the $P^{OX}/P^N$ couple is ca. −260 mV[89], suggesting a positive shift of nearly 1 volt from the potential of "normal" $Fe_4S_4$ clusters. Although this difference is larger than has been observed for other Fe-S proteins, it is still within the range that can be accounted for theoretically by variations in solvation and local dielectric constant[188].

Because of the extremely negative potentials associated with the 3-/4-couple, the fully reduced cluster, presumably isoelectronic with the $P^N$ state, has only been observed by stopped-flow reduction of $[Fe_4S_4(SCH_2CH_2CO_2)_4]^{6-}$ with chromous-EDTA complex[90], producing the highly unstable 8- ion. In contrast, the $[Fe_4S_4(SR)_4]^{3-}$ ions can be obtained

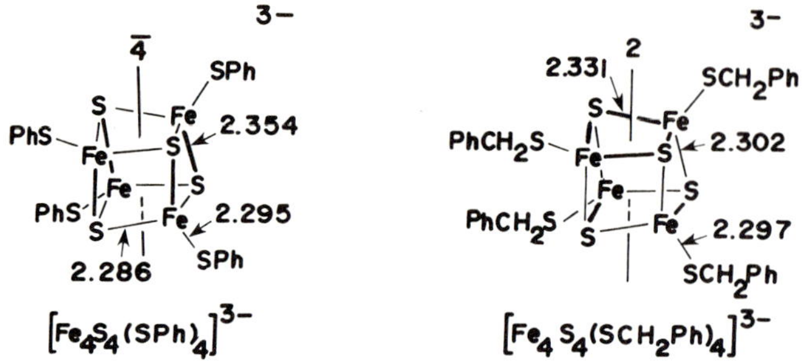


**Fig. 10.** Schematic drawing showing the distortions of the $Fe_4S_4^+$ core in: (*left*) $(Et_3MeN)_3$-$[Fe_4S_4(SPh)_4]$; (*right*) $(Et_4N)_3[Fe_4S_4(SCH_2Ph)_4]$. Reproduced with permission from Ref.[133]; copyright by the American Chemical Society

in pure crystalline form[189, 190], despite their extreme sensitivity to oxidation, and their proton magnetic resonance spectra[129], two crystal structure determinations (R = Ph[119], R = $CH_2Ph$[120]), and a variety of electronic properties[119, 133] have been reported. Some of these results appear to be directly relevant to the $P^{OX}$ state. Thus, two different structures have observed for the $Fe_4S_4$ core (see Fig. 10) in the structure determinations reported, and these different structures result in the observation of distinct electronic properties in the solid state. $(Et_3MeN)_3[Fe_4S_4(SPh)_4]$ contains a tetragonally *elongated* (by about 0.08 Å) $Fe_4S_4$ core[119], as opposed to the tetragonal compression observed for all $[Fe_4S_4(SR)_4]^{2-}$ structures to date[4]. In contrast, $(Et_4N)_3[Fe_4S_4(SCH_2Ph)_4]$ exhibits a non-tetragonal elongation of the $Fe_4S_4$ core along a twofold axis[120].

These structural differences are reflected in Mössbauer and EPR spectra and magnetic susceptibility and magnetization measurements on solid trianions[119, 133], which suggest the existence of two structural classes for a variety of cations and thiolates. Measurements on frozen solutions are consistent in all cases with the $[Fe_4S_4(SR)_4]^{3-}$ ions possessing solution structures closely related to that of $(Et_3MeN)_3[Fe_4S_4(SPh)_4]$, i.e., a tetragonally elongated $Fe_4S_4$ core[133]. This, the apparently more stable structure, has magnetic properties and EPR spectra typical of an axially distorted S = 1/2 unit[119, 133], entirely consistent with the properties of reduced 4 Fe-4 S ferredoxins[9–12]. Mössbauer spectra in zero applied field exhibit two overlapping quadrupole doublets in approximately 1 : 1 intensity ratios, while in high applied magnetic fields ($H_O$ = 80 kOe) two pairs of lines with opposite magnetic field dependences are observed, indicating the presence of two magnetically nonequivalent sites. Solid state measurements on the nontetragonally distorted trianion salts show only a broad quadrupole doublet in the zero field Mössbauer spectra, and smaller magnetic splitting (~ 40% of those observed for the tetragonally distorted trianions). The EPR spectra of these nontetragonal trianions are very complex, with features observed at apparent g values from 6 to 1.5, suggesting an S > 1/2 ground state. Magnetic susceptibility and magnetization measurements are consistent with this, and suggest a predominantly S = 3/2 ground state at low temperatures for the nontetragonal trianion structure[133].

The above results suggest that relatively minor crystal packing forces are sufficient to shift the $[Fe_4S_4(SR)_4]^{3-}$ ion from the more stable tetragonal S = 1/2 ground state structure to a nontetragonal S = 3/2 ground state structure. In fact, of nine different trianion salts examined, the majority exhibit the non-tetragonal structure in the solid state[133], although all apparently revert to the tetragonally elongated structure in solution. Although several theoretical calculations on these systems have appeared[126, 191, 192], none has suggested the presence of $S > 1/2$ states of energy comparable to the $S = 1/2$ ground state observed in the reduced ferredoxins. It would therefore seem to be premature to suggest that the tetragonally elongated and the nontetragonal twofold elongated structures are the only possibilities for $[Fe_4S_4(SR)_4]^{3-}$ ions. The electronic consequences of an imposed threefold distortion, such as one that would differentiate three of the Fe sites from the fourth, are of particular interest in regard to the $P^{OX}$ state; in view of the above, an $S = 5/2$ ground state, as has been suggested for the $P^{OX}$ clusters[89], does not seem at all farfetched.

## *B. $Fe_4S_4$ Clusters with Non-Thiolate Ligands*

Except for halide[142, 176] and benzeneselenate ($PhSe^-$)[193], the chemistry of tetranuclear Fe-S clusters with non-thiolate ligands has not been explored extensively. In particular, such complexes of potential physiological relevance are limited to acetate[36] and phenolate[141, 145] ligands, potential models for aspartate or glutamate and tyrosinate ligation, respectively. Hydrolysis of the $[Fe_4S_4]^{2+}$ core has been shown to proceed via coordination of $H_2O$ or $OH^-$ to iron[194], but these oxygen-ligated species are present only as transient intermediates under the conditions employed and no spectroscopic properties have been reported.

Of the oxygen-ligated $Fe_4S_4$ clusters for which spectroscopic data are available, $[Fe_4S_4(OAc)_4]^{2-}$ (**15**) has only been generated and examined in solution[36], while $[Fe_4S_4(OPh)_4]^{2-}$ (**16**)[141], and analogous complexes with other phenolate ligands[145] have been obtained as analytically pure crystalline materials. An X-ray structure determination on $(Et_4N)_2[Fe_4S_4(OPh)_4]$ has recently been completed[145]; the structure contains a normal tetragonally compressed $[Fe_4S_4]^{2+}$ core with relatively short (1.865 Å) Fe-O bonds.

Compared to $[Fe_4S_4(SPh)_4]^{2-}$, the results of oxygen ligation may be summarized as follows:

*(i)* both acetate and phenolate appear to be considerably more labile than thiolate ligands;
*(ii)* optical spectra are strongly blue-shifted for **15** and less so (20–40 nm) for **16**;
*(iii)* first and second reduction potentials are shifted to positive and to negative potentials by *ca.* 100 mV for **15** and **16**, respectively;
*(iv)* room temperature magnetic properties for **16** are essentially unchanged;
*(v)* isotropic shifts in the proton magnetic resonance are increased by a factor of ~2 for **16**;
*(vi)* the $^{57}Fe$ isomer shift in the Mössbauer spectrum increases by ~0.15 mm/s for **16** to 0.50 mm/s vs. metallic Fe at room temperature, consistent with increased ferrous character;
*(vii)* the quadrupole splitting for **16** increases only slightly, from 1.10 to 1.21 mm/s.

The results for **15** are similar to those observed upon halide substitution[36, 142], consistent with the trend expected upon replacing thiolate by more electronegative ligands. In contrast, the results for **16**, particularly the increased isotropic shifts in the NMR spectrum, the increase in $^{57}Fe$ chemical shift, and the negative shift in reduction potentials, are consistent with a relatively covalent Fe-O interaction.

Although the model for the P clusters presented earlier (Fig. 3) showed three tyrosyl (or aspartate or glutamate) and one cysteinyl ligand to the $Fe_4S_4$ core, obtaining data on such mixed-ligand complexes in synthetic systems is hampered by the lability of terminal ligands and the tendency to yield a statistical distribution of all possible mixed ligand complexes when a limiting amount of a second ligand is added. This is well documented for thiol substitution[140)] and has been demonstrated for mixed thiolate-acetate clusters as well[36)]. Consequently, it has not yet proven possible to prepare pure materials in which ligation at three Fe sites differs from that at the fourth. Nonetheless, it is clear from the results obtained to date that the presence of such a coordination environment in the P clusters, which would be significantly easier to accomplish with a relatively rigid protein matrix, is not inconsistent with the available data. Should this be the case, one would expect carboxylate or phenoxide ligation to result in a substitutionally more labile $Fe_4S_4$ center and in appreciable positive or negative shifts in reduction potentials, respectively. The discrepancy between the expected 3-/4-potential and that observed for the $P^{OX}/P^{N}$ couple argues in favor of a more electronegative ligand set, such as carboxylate, although nitrogenous ligands (imidazole or amine) cannot be ruled out.

### *C. Summary*

The P clusters appear to be unusual $Fe_4S_4$ units with anomalously positive potentials for the 3-/4-couple. In addition, the splitting of the Fe sites into sets of three and one within each cluster suggests that a trigonal perturbation is imposed upon the $Fe_4S_4$ center by the protein. Whether this perturbation is simply a protein-induced geometrical distortion, the addition of one or more extra ligands to three vertices of an $Fe_4S_4$ cube, or the replacement of three cysteinyl ligands by non-thiolate ligands is at present an open question, as is the origin of the marked positive shift in reduction potential.

## V. Conclusions

The results discussed above demonstrate that accurate synthetic analogs of both the FeMo-cofactor and of the P clusters have yet to be achieved. In spite of this, we now know vastly more about the properties of synthetic Mo-Fe-S and Fe-S clusters of potential relevance to the prosthetic groups of nitrogenase than we did only four years ago, when the first Mo-Fe-S double cubane clusters were prepared. The results obtained to date can be viewed as merely the first steps in accumulating the body of knowledge necessary for recognizing those features of the enzyme's prothetic groups that are essential for catalysis. Because physical studies on the enzyme have not yet allowed an unambiguous description of the composition and structures of the metal-sulfur chromophores, synthetic efforts in this area continue to be guided by intuition and serendipity as well as hard chemical facts. Of particular importance for furture work would seem to be incorporation of oxygen functionalities, both external to and within the cluster core, and the development of synthetic routes to new types of Mo-Fe-S structures, possibly utilizing starting materials other than tetrathiomolybdate.

This article has centered on the preparation of structural models, the accuracy of which may be assessed by comparison of corresponding physical properties of the model and the enzyme. Essentially no mechanistic details have been discussed, although a great deal of work has been performed[1, 195], particularly on reactions of $N_2$ coordinated to low valent Mo complexes, and mechanisms based on stepwise protonation and reduction of Mo-$N_2$ complexes have been proposed[55, 195]. The first phase of research, namely, preparation and structural and physiochemical characterization, has now largely been accomplished for the Mo-Fe-S complexes prepared thus far, although synthetic efforts aimed at new types of Mo-Fe-S clusters continue unabated. The second phase, investigation of the reaction chemistry of these complexes with small molecules, is just beginning. Only when more data are available will we be able to ascertain whether these polynuclear Mo-Fe-S clusters exhibit reactivity analogous to the better-studied mononuclear Mo complexes. Whether the P clusters prove to function simply as electron reservoirs or whether they participate directly in reduction of one or more substrates, it seems likely that their function is intimately related to the unusual perturbation imposed upon the 4 Fe-4 S unit by the enzyme.

Finally, one of the major functions of research on synthetic systems such as those discussed above is to suggest new experiments aimed at elucidating the mechanism of catalysis by the enzyme, as well as to provide information on the intrinsic properties of the metal-sulfur centers in the absence of the protein environment.

*Acknowledgements.* Research in the author's laboratory has been supported by the National Science Foundation (Grant # CHE-7715990) and by the United States Department of Agriculture Science and Education Administration's Competitive Research Grants Office (5901-0410-8-0175-0). I thank B. K. Burgess, R. H. Holm, W. E. Newton, S. Otsuka, B. E. Smith, and J. A. Zubieta for communication of results prior to publication, and W. H. Orme-Johnson for stimulating discussions.

## VI. References

1. Chatt, J., Dilworth, J. R., Richards, R. L.: Chem. Rev. *78*, 589 (1978)
2. Stiefel, E. I.: Progr. Inorg. Chem. *22*, 1 (1977)
3. Holm, R. H.: Accts. Chem. Res. *10*, 427 (1977)
4. Holm, R. H., Ibers, J. A.: Synthetic analogues of the active sites of iron-sulfur proteins, in: Iron-Sulfur Proteins, V. 3 (ed.) Lovenberg, W., p. 205, New York, Academic Press 1977
5. Averill, B. A., Orme-Johnson, W. H.: Iron-sulfur proteins and their synthetic analogs, in: Metal Ions in Biological Systems, V. 7, Iron in Model and Natural Compounds (ed.) Sigel, H., p. 127, New York, Marcel Dekker, 1978
6. Coughlan, M. P. (ed.): Molybdenum and Molybdenum-Containing Enzymes, Oxford – New York – Toronto – Sydney – Paris – Frankfurt, Pergamon Press 1980
7. Newton, W. E., Otsuka, S. (eds.): Molybdenum Chemistry of Biological Significance, New York – London, Plenum Press 1980
8. Orme-Johnson, W. H. et al.: Molybdenum in enzymes, in Bioinorganic Chemistry-II (ed.) Ramond, K., p. 389, ACS Adv. Chem. Ser. *162* (1977)
9. Palmer, G.: Iron-sulfur proteins, in: The Enzymes 3rd Ed., V. XII (ed.) Boyer, P., p. 1, New York, Academic Press 1975[3]
10. Lovenberg, W. (ed.): Iron-Sulfur Proteins, V. 1, New York, Academic Press 1973
11. Lovenberg, W. (ed.): Iron-Sulfur Proteins, V. 2, New York, Academic Press 1973
12. Lovenberg, W. (ed.): Iron-Sulfur Proteins, V. 3, New York, Academic Pres 1977
13. Mortenson, L. E., Thorneley, R. N. F.: Ann. Rev.Biochem. *48*, 387 (1979)

14. Orme-Johnson, W. H. et al.: Components and pathways in biological nitrogen fixation, in: Recent Development in Nitrogen Fixation, (ed.) Newton, W. E., Postgate, J., Rodriguez-Barrueco, C., p. 131, New York – London, Academic Press 1977
15. Brill, W. J.: Microbiol. Rev. *44*, 449 (1980)
16. Robson, R. L., Postgate,J. R.: Ann. Rev. Microbiol. *34*, 183 (1980)
17. Hardy, R. W. F., Bottomley, F., Burns, R. C. (eds.): A Treatise on Dinitrogen Fixation, Sections I and II, New York, Wiley-Interscience 1979
18. Hardy, R. W., Burns, R. C., (eds): Nitrogen Fixation in Bacteria and Higher Plants, Heidelberg – Berlin – New York, Springer-Verlag 1975
19. Müller, A., Newton, W. E. (eds.): Nitrogen Fixation: Chemical/Biochemical/Genetics Interface, New York, Plenum Press 1981
20. Newton, W. E., Nyman, C. J. (eds.) Proc. 1st Intl. Symp. Nitrogen Fixation, Pullman, Washington State University Press 1976
21. Newton, W. E., Postgate, J., Rodriguez-Barrueco, C. (eds.): Recent Developments in Nitrogen Fixation, New York, Academic Press 1977
22. Newton, W. E., Orme-Johnson, W. H. (eds.): Nitrogen Fixation, Baltimore, University Park Press 1980
23. Gibson, A. H., Newton, W. E.: Current Perspectives in Nitrogen Fixation, Canberra, Australian Academy of Science 1981
24. Hill, H. A. O.: Chem. in Brit. *12*, 119 (1976)
25. Ibers, J. A., Holm, R. H.: Science *209*, 223 (1980)
26. Orme-Johnson, W. H., Davis, L. C.: Current topics and problems in the enzymology of nitrogenase, in Ref. 12, p. 15
27. Ludden, P. W., Burris, R. H.: Science *194*, 424 (1976)
28. Ludden, P. W., Burris, R. H.: Proc. Natl, Acad. Sci. U.S.A. *76*, 6201 (1979)
29. Hageman, R. V., Burris, R. H.: Proc. Natl. Acad. Sci. U.S.A. *75*, 2699 (1978)
30. Tso, M.-Y. W., Burris, R. H.: Biochim. Biophys. Acta. *309*, 263 (1973)
31. Smith, B. E., Lowe, D. J., Bray, R. C.: Biochem. J. *135*, 331 (1973)
32. Zumft, W. G., Palmer, G., Mortenson, L. E.: Biochim. Biophys. Acta. *292*, 413 (1973)
33. Gillum, W. O. et al.: J. Amer. Chem. Soc. *99*, 584 (1977)
34. Hageman, R. V., Orme-Johnson, W. H., Burris, R. H.: Biochemistry *19*, 2333 (1980)
35. Zumft, W. G., Mortenson, L. E., Palmer, G.: Eur. J. Biochem. *46*, 525 (1974)
36. Johnson, R. W., Holm, R. H.: J. Amer. Chem. Soc. *100*, 5338 (1978)
37. Kennedy, C., Eady, R. R., Kondorosi, E.: Biochem. J. *155*, 383 (1976)
38. Burgess, B. K., Jacobs, D. R., Stiefel, E. I.: Biochim. Biophys. Acta *614*, 196 (1980)
39. Zimmerman, R. et al.: Biochim. Biophys. Acta *537*, 185 (1978)
40. Huynh, B. H. et al.: Biochim. Biophys. Acta *623*, 124 (1980)
41. Tonsager, S., Averill, B. A.: Anal. Biochem. *102*, 13 (1980)
42. Münck, E. et al.: Biochim. Biophys. Acta *400*, 32 (1975)
43. Palmer, G. et al.: Arch. Biochem. Biophys. *153*, 325 (1972)
44. Smith, B. E., Lang, G.: Biochem. J. *137*, 169 (1974)
45. Jarrett, H. S.: J. Chem. Phys. *27*, 1298 (1957)
46. Averill, B. A., Orme-Johnson, W. H.: Inorg. Chem. *19*, 1702 (1980)
47. Orme-Johnson, W. H. et al: Spectroscopic and chemical evidence of the nature and role of metal centers in nitrogenase and nitrate reductase, in: Ref. 7, p. 85
48. Huynh, B. H., Münck, E., Orme-Johnson, W. H.: Biochim. Biophys. Acta *527*, 192 (1979)
49. Smith, B. E. et al.: Biochem. J. *191*, 449 (1980)
50. Nagatani, H. H., Shah, V. K., Brill, W. J.: J. Bacteriol. *120*, 697 (1974)
51. Shah, V. K., Brill, W. J.: Proc. Natl. Acad. Sci. U.S.A. *74*, 3249 (1977)
52. Pienkos, P. T., Shah, V. K., Brill, W. J.: Proc. Natl. Acad. Sci. U.S.A. *74*, 5468 (1977)
53. Johnson, J. L., Hainline, B. E., Rajagopalan, K. V.: J. Biol. Chem. *255*, 1783 (1980)
54. Shah, V. K., Chisnell, J. R., Brill, W. J.: Biochem. Biophys. Res. Commun. *81*, 232 (1978)
55. Weathers, B. J. et al.: J. Amer. Chem. Soc. *101*, 925 (1979)
56. Schrauzer, G. N. et al.: J. Amer. Chem. Soc. *98*, 2815 (1976)
57. McMillan, R. S. et al.: J. Inorg. Biochem. *11*, 213 (1979)
58. Burgess, B. K.: Iron-molybdenum cofactor and its complementary protein from mutant organisms, in Ref. 19

59. Burgess, B. K., Newton, W. E.: in: Ref. 19
60. Smith, B. E.: presented at the International Symposium on Molybdenum Chemistry of Biological Significance, Kyoto, 1979
61. Newton, W. E., Burgess, B. K., and Stiefel, E. I.: Chemical properties of the Fe-Mo cofactor from nitrogenase, in: Ref. 7, p. 191
62. Smith, B. E.: Studies on the iron-molybdenum cofactor from the nitrogenase Mo-Fe protein of Klebsiella pneumoniae, in: Ref. 7, p. 179
63. Levchenko, L. A. et al.: Biochem. Biophys. Res. Commun. *96*, 1384 (1980)
64. Shah, V. K., Brill, W. J.: Proc. Natl. Acad. Sci. U.S.A. *78*, 3438 (1981)
65. Burgess, B. K., Stiefel, E. I., Newton, W. E.: J. Biol. Chem. *255*, 353 (1980)
66. Rawlings, J. et al.: J. Biol. Chem. *253*, 1001 (1978)
67. Burgess, B. K. et al.: Iron-molybdenum cofactor and its complementary protein from *Azotobacter vinelandii* UW 45, in Ref. 23, p. 71
68. Cramer, S. P. et al.: J. Amer. Chem. Soc. *98*, 1287 (1976)
69. Cramer, S. P. et al.: J. Amer. Chem. Soc. *100*, 3398 (1978)
70. Cramer, S. P. et al.: J. Amer. Chem. Soc. *100*, 3814 (1978)
71. Wolff, T. E. et al.: J. Amer. Chem. Soc. *100*, 4630 (1978)
72. Smith, J. P. et al.: Investigation of molybdenum-substrate interactions with X-ray absorption spectroscopy in Ref. 22, V. 1, p. 283
73. Antonio, M. R. et al.: submitted for publication
74. Zumft, W. G.: Eur. J. Biochem. *91*, 345 (1978)
75. Harmer, M. A., Sykes, A. G.: Inorg. Chem. *19*, 2881 (1980)
76. Wolff, T. E. et al.: J. Amer. Chem. Soc. *101*, 4140 (1979)
77. Teo, B.-K., Averill, B. A.: Biochem. Biophys. Res.Commun. *88*, 1454 (1979)
78. Ghosh, D. et al.: J. Biol. Chem. *256*, 4185 (1981)
79. Huynh, B. H. et al.: J. Biol. Chem. *255*, 3242 (1980)
80. Kent, T. A., Huynh, B. H., Münck, E.: Proc. Natl. Acad. Sci. U.S.A. *77*, 6574 (1980)
81. Lu, J.: Composite "string bag" cluster model for the active center of nitrogenase, in: Ref. 22, v. 1, p. 343
82. Chu, C. T.-W., Dahl, L. F.: Inorg. Chem. *16*, 3245 (1977)
83. Wong, G. B. et al.: J. Amer. Chem. Soc. *101*, 3078 (1979)
84. Kurtz, D. M., Jr. et al.: Proc. Natl. Acad. Sci. U.S.A. *76*, 4986 (1979)
85. Davis, L. C. et al.: Biochemistry *18*, 4860 (1979)
86. Stephens, P. J. et al.: Proc. Natl. Acad. Sci. U.S.A. *76*, 2585 (1979)
87. Johnson, M. K. et al.: Biochim. Biophys. Acta *671*, 61 (1981)
88. Rao, K. K. et al.: Biochem. J. *129*, 1063 (1972)
89. Orme-Johnson, W. H. et al.: Nitrogenase: prosthetic groups and their reactivities, in Ref. 23, p. 79
90. Henderson, R. A., Sykes, A. G.: Inorg. Chem. *19*, 3103 (1980)
91. Cotton, F. A., Wilkinson, G.: Advanced Inorganic Chemistry: a Comprehensive Text, p. 755, 759, New York, John Wiley and Sons 1980[4]
92. Christou, G., Garner, C. D., Mabbs, F. E.: Inorg. Chim. Acta *28*, L 189 (1978)
93. Diemann, E., Müller, A.: Coord. Chem. Rev. *10*, 79 (1973)
94. Dias, A. R., Green, M. L. H.: J. Chem. Soc A, 2807 (1971)
95. Cameron, T. S., Prout, C. K.: Acta Crystallogr., *B 28*, 453 (1972)
96. Simhon, E. D. et al.: J. Amer. Chem. Soc. *103*, 1218 (1981)
97. Müller, A., Nolte, W. O., Krebs, B.: Angew. Chem. Int. Ed. Engl. *17*, 279 (1978)
98. Müller, A. et al.: Angew. Chem. Int. Ed. Engl. *17*, 535 (1978)
99. Millar, M., Koch, S. A.: Abstr. 178th American Chem. Soc. Meet., INOR-242, Washington, 1979
100. Bunzey, G. et al.: J. Amer. Chem. Soc. *99*, 4168 (1977)
101. Bunzey, G., Enemark, J. H.: Inorg. chem. *17*, 682 (1978)
102. Miller, K. F. et al.: J. Amer. Chem. Soc. *102*, 5102 (1980)
103. Hyde, J., Magin, L., Zubieta, J.: J. Chem. Soc. Chem. Comm. 204 (1980)
104. Otsuka, S. et al.: J. Amer. Chem. Soc. *103*, 3011 (1981)
105. Kamata, M. et al.: J. Amer. Chem. Soc. *103*, 3572 (1981)
106. Nagatani, H. H., Brill, W. J.: Biochim. Biophys. Acta *362*, 160 (1974)

107. Wolff, T. E. et al.: J. Amer. Chem. Soc. *101*, 5454 (1979)
108. Wolff, T. E. et al.: J. Amer. Chem. Soc. *102*, 4694 (1979)
109. Christou, G. et al.: J. Chem. Soc. Chem. Comm. 740 (1978)
110. Acott, S. R. et al.: Inorg. Chim. Acta *35*, L337 (1979)
111. Christou, G., Garner, C. D.: J. Chem. Soc., Dalton Trans. 2354 (1980)
112. Christou, G. et al.: J. Chem. Soc. Chem. Comm. 503 (1979)
113. Wolff, T. E. et al.: Inorg. Chem. *19*, 430 (1980)
114. Christou, G. et al.: J. Chem. Soc. Chem. Comm. 91 (1979)
115. Garner, C. E. et al.: Studies of iron-molydenum-sulphur clusters, in: Ref. 23, p. 40
116. Wolff, T. E., Berg, J. M., Holm, R. H.: Inorg. Chem. *20*, 174 (1981)
117. Christou, G. et al.: J. Inorg. Biochem. *11*, 349 (1979)
118. Averill, B. A. et al.: J. Amer. Chem. Soc. *95*, 3523 (1973)
119. Laskowski, E. J. et al.: J. Amer. Chem. Soc. *100*, 5322 (1978)
120. Berg, J. M., Hodgson, K. O., Holm, R. H.: J. Amer. Chem. Soc. *101*, 4586 (1979)
121. Schäfer, H., Schäfer, G., Weiss, A.: Z. Naturforsch. *19b*, 76 (1964)
122. Kozmin, P. A., Popova, Z. V.: J. Struct. Chem. *12*, 81 (1971)
123. Saillant, R. et al.: Inorg. Chem. *10*, 1453 (1971)
124. DePamphilis, B. V. et al.: J. Amer. Chem. Soc. *96*, 4159 (1974)
125. Hill, C. L. et al.: J. Amer. Chem. Soc. *99*, 2549 (1977)
126. Yang, C. Y. et al.: J. Amer. Chem. Soc. *97*, 6596 (1975)
127. Averill, B. A., Bale, J. R., Orme-Johnson, W. H.: J. Amer. Chem. Soc. *100*, 3034 (1978)
128. Holm, R. H. et al.: J. Amer. Chem. Soc. *96*, 2109 (1974)
129. Reynolds, J. G., Laskowski, E. J., Holm, R. H.: J. Amer. Chem. Soc. *100*, 5315 (1978)
130. Christou, G. et al.: J. Chem. Soc. Dalton Trans. 2363 (1980)
131. Cammack, R., Dickson, D. P. E., Johnson, C. E.: Evidence from Mössbauer spectroscopy and magnetic resonance on the active centers of the iron-sulfur proteins, in: Ref. 12, p. 283
132. Frankel, R. B., Averill, B. A., Holm, R. H.: J. Phys. (Paris) *35*, C6–107 (1974)
133. Laskowski, E. J. et al.: J. Amer. Chem. Soc. *101*, 6562 (1979)
134. Christou, G. et al.: Inorg. Nucl. Chem. Lett. *17*, 137 (1981)
135. Gillum, W. O. et al.: Inorg. Chem. *15*, 1095 (1978)
136. Blum, H. et al.: Biochem. Biophys. Res. Commun. *77*, 650 (1977)
137. Norman, J. G., Jr., Ryan, P. B., Noodleman, L.: J. Amer. Chem. Soc. *102*, 4279 (1980)
138. Christou, G., Hageman, R. V., Holm. R. H.: J. Amer. Chem. Soc. *102*, 7600 (1980)
139. Adams, M. W. W. et al.: Biochem. Biophys. Acta *589*, 1 (1980)
140. Que, L. Jr. et al.: J. Amer. Chem. Soc. *96*, 4168 (1974)
141. Cleland, W. E., Averill, B. A.: Inorg. Chim. Acta *56*, L9 (1981)
142. Wong, G. B., Bobrik, M. A., Holm, R. H.: Inorg. Chem. *17*, 578 (1978)
143. Christou, G., Garner, C. D.: J. Chem. Soc. Chem. Comm. 613 (1980)
144. Palermo, R. E., Power, P. P., Holm, R. H.: Inorg. Chem. *21*, 173 (1982)
145. Cleland, W. E. et al.: to be submitted for publication
146. Otsuka, S., Kamata, M.: Molybdo-iron mixed cluster as Fe-Mo cofactor model, in: Ref. 7, p. 229
147. Avdeef, A. et al.: J. Amer. Chem. Soc. *100*, 5362 (1978)
148. Benemann, J. R. et al.: FEBS Lett. *29*, 219 (1973)
149. Wallace, T. J., Mahon, J. J.: J. Amer. Chem. Soc. *86*, 4099 (1964)
150. Müller, A., Diemann, E.: J. Chem. Soc. Chem. Comm. 65 (1971)
151. Müller, A., Ahlborn, E., Heinsen, H. H.: Z. Anorg. Allg. Chem. *386*, 102 (1971)
152. Müller, A., Diemann, E., Heinsen, H. H.: Chem. Ber. *104*, 975 (1971)
153. Callahan, K. P., Piliero, P. A.: J. Chem. Soc. Chem. Commun. 13 (1979)
154. Sotofte, I.: Acta Chim. Scand. *A30*, 157 (1976)
155. Müller, A., Mohan, N., Bögge, H.: Z. Naturforsch. *33b*, 978 (1978)
156. Paulat-Böschen, I. et al.: Inorg. Chem. *17*, 1440 (1978)
157. Müller, A., Sarkar, S.: Angew. Chem. Int. Ed. Engl. *16*, 705 (1977)
158. Coucouvanis, D.: Accts. Chem. Res. *14*, 201 (1981)
159. Coucouvanis, D. et al.: J. Chem. Soc. Chem. Comm. 361 (1979)
160. Tieckelmann, R. H. et al.: J. Amer. Chem. Soc. *102*, 5550 (1980)
161. Müller, A., Tolle, H.-G., Bögge, H.: Z. Anorg. Allg. Chem. *471*, 115 (1980)

162. Müller, A. et al.: Inorg. Chim. Acta *46*, L 121 (1980)
163. Müller, A. et al.: Angew. Chem. Int. Ed. Engl. *19*, 654 (1980)
164. Silvis, H. C., Averill, B. A.: Inorg. Chim. Acta *54*, L 57 (1981)
165. Coucouvanis, D. et al.: Inorg. Chim. Acta *53*, L 135 (1981)
166. Coucouvanis, D. et al.: J. Amer. Chem. Soc. *102*, 1730 (1980)
167. Coucouvanis, D. et al.: J. Amer. Chem. Soc. *102*, 1732 (1980)
168. Müller, A. et al.: Z. Naturforsch. *35 b*, 1592 (1980)
169. Tieckelmann, R. H., Averill, B. A.: Inorg. Chim. Acta *46*, L 35 (1980)
170. Tieckelmann, R. H., et al.: to be submitted for publication
171. McDonald, J. W., Friesen, G. D., Newton, W. E.: Inorg. Chim. Acta L 79, 46 (1980)
172. Coucouvanis, D., Simhon, E. D., Baenziger, N. C.: J. Amer. Chem. Soc. *102*, 6644 (1980)
173. Dahlstrom, P. L., Kumar, S., Zubieta, J.: J. Chem. Soc. Chem. Comm. 411 (1981)
174. Silvis, H. C. et al.: to be submitted for publication
175. Mayerle, J. J. et al.: J. Amer. Chem. Soc. *97*, 1032 (1975)
176. Bobrik, M. A., Hodgson, K. O., Holm, R. H.: Inorg. Chem. *16*, 1851 (1977)
177. Silvis, H. C.: Ph. D. thesis, Michigan State University, 1981
178. Hush, N. S.: Progr. Inorg. Chem. *8*, 391 (1967)
179. Schmidt, K. H., Müller, A.: Coord. Chem. Rev. *14*, 115 (1974)
180. Müller, A.: Molybdenum-sulfur chemistry related to nitrogen fixation, in: Ref. 23, p. 44
181. Newton, W. E. et al.: Molybdenum-iron-sulfur complexes and their relevance to the molybdenum site of nitrogenase, in: Ref. 23, p. 30
182. Müller, A. et al.: Inorg. Chim. Acta *44*, L 33 (1980)
183. Stremple, P., Baenziger, N. C., Coucouvanis, D.: J. Amer. Chem. Soc. *103*, 4601 (1981)
184. Huang, W. et al.: Attempted synthesis of "Series G" model compounds for the active center of nitrogenase, in Ref. 23, p. 346
185. Lu, J.: Evidence for the "string-bag" structure as a basic structural unit for the active center of nitrogenase and an attempted synthesis of "string-bag" model compounds, in Ref. 23, p. 50
186. Tsai, K. R. et al.: Studies on the mechanism and chemical modelling of nitrogenase catalysis, in: Ref. 23, p. 344
187. Lu, J.: Attempted synthesis of "Series F" model compounds for the active center of nitrogenase, in Ref. 23, p. 345
188. Kassner, R. J., Yang, W.: J. Amer. Chem. Soc. *99*, 4351 (1977)
189. Lane, R. W. et al.: J. Amer. Chem. Soc. *99*, 2350 (1977)
190. Cambray, J. et al.: Inorg. Chem. *16*, 2565 (1977)
191. Geurts, P. J. M. et al.: Chem. Phys. *46*, 133 (1980)
192. Thomson, A. J.: J. Chem. Soc. Dalton Trans. 1180 (1981)
193. Bobrik, M. A. et al.: Inorg. Chem. *17*, 1402 (1978)
194. Job, R. C., Bruice, T. C.: Proc. Natl. Acad. Sci. U.S.A. *72*, 2478 (1975)
195. Chatt, J.: Chemtech. *11*, 152 (1981)
196. Mortenson, L. E., Chen, J.-S.: Hydrogenase, in: Microbial Iron Metabolism (ed.) Neilands, J. B., p. 231, New York, Academic Press 1974
197. Gillum, W. O. et al.: J. Amer. Chem. Soc. *99*, 584 (1977)
198. Erbes, D. L., Burris, R. H., Orme-Johnson, W. H.: Proc. Natl. Acad. Sci. U.S.A. *72*, 4795 (1975)
199. Averill, B. A., Orme-Johnson, W. H.: J. Amer. Chem. Soc. *100*, 5234 (1978)

# The Biochemistry of Vanadium

**N. Dennis Chasteen**

Department of Chemistry, University of New Hampshire, Durham, NH 03824, USA

Vanadium is an essential element. It is found widely distributed in minor and trace amounts throughout the lithosphere and biosphere. Vanadium is present in all mammalian tissues at concentrations of about 10 μM or less. Numerous physiological effects of this element are known at both the organismic and cellular levels, and in some instances these effects can be explained in terms of its biochemistry. Vanadate(V) and/or vanadyl(IV) ions are very potent inhibitors ($K_I \lesssim 10^{-6}$ M) of certain phosphatases, ATPases, phosphotransferases, nucleases, and kinases, among others. Vanadate(V) activates adenylate kinase. These ions have been quite useful as kinetic and spectroscopic (EPR) probes of complex cellular processes. Vanadium is an insulin mimetic agent. The ability of vanadium at low concentrations to influence enzymes is probably important to its function in vivo. Much of the biochemistry of vanadium is intricately related to the redox and coordination chemistry of this element. In particular, the parallel between the chemistries of vanadates and phosphates can account for the observed inhibition of various phosphohydrolases. However, there is ample evidence that vanadium in the body exists primarily as the vanadyl(IV) ion, $VO^{2+}$, complexed to proteins and other cellular components; several biological reducing agents can convert vanadium(V) to this form. Serum transferrin appears to be involved in the metabolism of vanadium, perhaps as the transport protein for $VO^{2+}$. Despite recent major advances in our understanding of the biochemistry of vanadium, the delineation of the physiological function(s) of this important element has remained an elusive goal.

## Abbreviations

| | |
|---|---|
| ADP | adenosine diphosphate |
| AMP | adenosine monophosphate |
| cAMP | cyclic adenosine monophosphate |
| ATP | adenosine triphosphate |
| $Ca^{2+}$-ATPase | calcium ion stimulated adenosine triphosphatase |
| DNA | deoxyribonucleic acid |
| ENDOR | electron-nuclear double resonance |
| EPR | electron paramagnetic resonance |
| ESE | electron spin echo |
| EXAFS | extended X-ray absorption fine structure |
| GSH | glutathione |
| GTP | guanosine triphosphate |
| GTP-γ-S | guanosine 5′-0-(3-thiotriphosphate) |
| Hepes | N-2-hydroxyethylpiperazine-N′-2-ethane sulfonic acid |
| NADH | nicotinamide adenine dinucleotide (reduced) |
| $Na^+$,$K^+$-ATPase | sodium and potassium ions stimulated adenoisine triphosphatase |
| NMR | nuclear magnetic resonance |
| $P_i$ | orthophosphate |
| $PP_i$ | pyrophosphate |
| PNPP | p-nitrophenylphosphate |
| RNA | ribonucleic acid |
| tris | tris(hydroxymethyl)aminomethane |
| UV | ultraviolet |
| VO($acac_2$(pn)) | vanadyl(IV)bisacetylacetonepropylenediimine |
| VOTPP | vanadyl(IV)tetraphenylporphyrin |
| ZnTPP | zinc(II)tetraphenylporphyrin |

## A. Introduction

Vanadium was first discovered in 1801 by Andrés Manuel del Rio, professor of mineralogy at the School of Mines Mexico[1]. He originally named the element panchromium after the varied colors of its salts, but later renamed it erythronium because of the red color generated upon heating. Unfortunately, del Rio lost confidence in his discovery, thinking that he had simply found the element chromium, which had recently been discovered by the Frenchman Fourcroy. Vanadium was rediscovered in 1831 by the Swede Nils Gabriel Sefström. Its present name is derived from Vanadis, the goddess of love and beauty of Norse mythology.

Vanadium is found widely distributed in rocks, soils, plants, animals, and to a lesser extent in waters where levels rarely exceed 0.1 ppb[2–4]. The concentration in seawater is but 50 nM[5]. As the result of diagenesis, particularly large accumulations of up to 1% are often found in coal and petroleum deposits. Such high levels pose problems in coal conversion, petroleum refining, and in the production of airborne vanadium as a byproduct of combustion of vanadium containing fuels[3, 6, 7]. The vanadium levels in the basaltic and igneous rocks of the earth's crust average 150 ppm, ranking 10th in abundance of the approximately 29 elements which may have a biological role[4, 8]. Commercially viable amounts of vanadium occur in titaniferous magnetite deposits containing approximately 7000 ppm of the element[4]. Neither elemental vanadium nor its compounds are found in a pure form in nature.

The distribution of vanadium in the surface environment can be traced to its tendency to be associated with organic matter. The oxovanadium(IV) ion, $VO^{2+}$, is strongly sequestered by humic and fulvic acids derived from soils and sediments[9–11]. The same may be true of the dioxovanadium(V) ion, $VO_2^+$. Orthovanadate(V), $HVO_4^{2-}$, is a more mobile form of vanadium. Vanadium in fruits, vegetables, and grains typically falls in the range 0.1 to 5 ppm on a dry weight basis[2–4, 8]. The element is often found concentrated in fats and oils.

Vanadium was long suspected to be an essential element. In the early 1970's, the essentiality of vanadium was confirmed with experiments employing laboratory animals[12, 13]. The normal growth and development of rats and chicks is retarded when these animals are raised on synthetic diets containing less than 10 ppb vanadium. The dietary requirement appears to be around 0.1 ppm although it may be considerably higher[4]. The levels necessary to cause symptoms of deficiency may be influenced by the other nutrients also present in the diet[14].

The physiological effects of vanadium are many. Adverse exposure is generally related to airborne vanadium since relatively little ingested vanadium is taken up in the gut. Toxicological studies have been reviewed[2, 3, 15]. Epidemiological research suggests an inverse correlation between cardiovascular disease and vanadium levels in water[3, 4]. It has been known for years that vanadium reduces the biosynthesis of cholesterol and the plasma levels of triglycerides in young subjects. Currently, the effects of vanadate(V) on the cardiovascular system is being intensively explored. Some studies suggest that vanadium enhances mineralization of teeth and bone in laboratory animals. There is evidence that vanadium may be involved in glucose metabolism, renal and bladder function, the sodium pump, muscle contractions and other physiological processes (vide infra). A link between hormonal action and vanadium levels appears to exist. Numerous enzymes are inhibited by vanadium ions or their complexes. Table 1 summarizes the various physiological and biochemical aspects of vanadium known to date. An overview of vanadium in biology and biochemistry and the molecular basis for some of these observations will be presented in this chapter.

The number of studies of the biochemistry and physiology of vanadium has grown enormously in recent years. Because of the volume of published material, it is not possible to cover all of the important contributions in this area. Our knowledge of vanadium in biology is presently very much in a state of flux. Those aspects of biochemistry of vanadium which currently are best understood will receive proportionately greater attention here. In addition, the chemical and spectroscopic properties of vanadium that are relevant to studies of the biochemistry of this element will also be briefly covered. Some reviews of limited scope have appeared[16–21].

**Table 1.** Aspects of the biochemistry and physiology of vanadium

| Property | Leading References |
|---|---|
| Widespread in nature | 4 |
| Essential element | 12, 13 |
| Dietary requirement perhaps near 0.1 ppm | 8, 14, 203 |
| Toxic effects observed at dietary levels of 25 ppm | 203 |
| Especially toxic as airborne vanadium | 2, 3, 15 |
| Deficiency impairs growth and reproduction | 3, 12, 13, 203 |
| Deficiency elevates erythrocyte and iron levels | 8, 203 |
| Suppresses cholesterol biosynthesis and reduces plasma triglycerides in vivo | 3, 4, 14, 15, 159, 203 |
| Positive and negative effects on strength of contraction of cardiovascular muscle | 130–152 |
| Possibly imparts protection against dental caries | 3, 203 |
| Exerts diuretic and natriuretic effects as vanadate | 19 |
| Possible link between manic depressive psychosis and vanadium levels in serum | 160–163 |
| Possible relationship between vanadium metabolism and hormonal effects in reproduction | 164–167 |
| Influence on the metabolism of molybdenum | 154–158 |
| Possible link between vanadium and iron metabolism | 186, 187 |
| Decreases intraocular pressure in eye | 169, 170 |
| Causes insulin like response in cells | 33, 102–105 |
| Increases cAMP and protein kinase levels in liver tissue | 123 |
| Binds to proteins as vanadyl(IV) and vanadate(V) | 16, 18, 21 |
| Inhibits numerous enzymes | Table 2 |
| Accumulates in the cap of the mushroom *Amanita muscaria* | 106–108 |
| Concentrates in vanadocyte blood cells of ascidians | 113 |

## B. Inorganic Chemistry of Vanadium

The chemistry of vanadium is characterized by multiple oxidation states (Fig. 1). The redox chemistry of this metal undoubtedly plays a role in its biochemistry. Of the four common oxidation states, only V(III), V(IV), and V(V) are important biologically, V(II) being too reducing to exist in any known organism. The best known example of the occurrence of V(III) is in the vanadocytes of the blood of tunicates[22)]; otherwise, vanadium is largely found in the +4 and +5 oxidation states, both of which are readily accessible under physiological conditions.

In acid, the predominent species are $V^{3+}$, $VO^{2+}$, and *cis*-$VO_2^+$ (Fig. 1). For simplicity, coordinated water molecules have been omitted from these formulas. As the pH is raised, hydrolysis takes place and a number of monomeric and oligomeric species are formed, only some of which are shown in Figure 1. Because of the multiple equilibria involved and the tendency for equilibrium to be attained slowly, the elucidation of all of the species present and their respective formation constants is a difficult task.

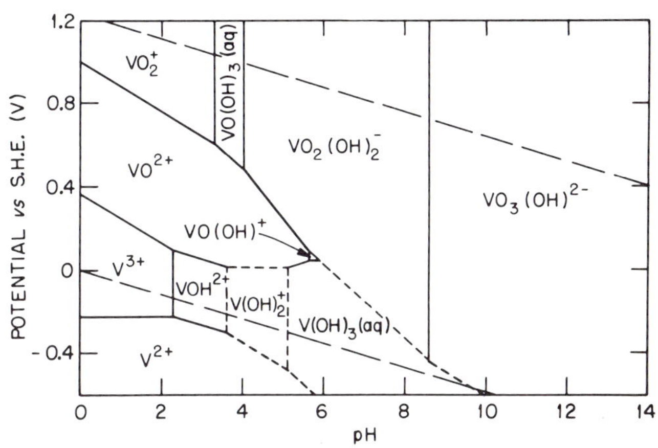


**Fig. 1.** Reduction potential E (referenced to the standard hydrogen electrode) versus pH for various species of vanadium. Boundary lines correspond to E, pH values where the species in adjacent regions are present in equal concentrations. The short dashed lines indicate uncertainty in the location of the boundary. The *upper* and *lower long dashed lines* correspond to the upper and lower limits of stability of water. Standard reduction potentials are given by the intersections of "horizontal" lines with the abscissa pH = 0. The half reactions are $O_2 + 4H^+ + 4e = 2H_2O$, $E° = 1.23V$; $VO_2^+ + 2H^+ + e = VO^{2+} + H_2O$, $E° = 1.0V$; $VO^{2+} + 2H^+ + e = V^{3+} + H_2O$, $E° = 0.36V$; $2H^+ + 2e = H_2$, $E° = 0.0V$; and $V^{3+} + e = V^{2+}$, $E° = -0.25V$. $V^{2+}$ is therefore a strong reductant. Air oxidation of $VO^{2+}$ presumably proceeds by the reaction $4VO^{2+} + O_2 + 2H_2O = 4VO_2^+ + 4H^+$, $E° = 0.23V$ which is favored at higher pH. Not all known species are represented on this diagram. Reproduced with permission from Ref. 30

### *1. Vanadium (V)*

In the physiological pH range 6 to 8 when the total vanadium concentration is less than 10 mM, the species present in appreciable amounts are the vanadates(V) $H_2VO_4^-$, $HVO_4^{2-}$, $HV_2O_7^{3-}$, and $V_3O_9^{3-}$ in which the metal is tetrahedrally coordinated[23, 24)]. In the presence of chelating agents such as EDTA, octahedral complexes containing the *cis*-$VO_2^+$ entity are formed[25]. There is a tendency for vanadium(V) to complex with glycols and catechols.

The inhibitory or activating effect of vanadium on many phosphohydrolases (*vide infra*) appears to largely, but not exclusively, reside in the +5 oxidation state. The similarity between the chemistries of the vanadates(V) and the phosphates is undoubtedly responsible for much of the biological activity of this metal. VO bond lengths in vanadates are around 0.17 nm compared to 0.152 nm for the P–O bond in orthophosphate. Vanadic acid $H_3VO_4$ ($pK_a$ = 3.5, 7.8, 12.5) is a weaker acid than phosphoric acid $H_3PO_4$ ($pK_a$ = 1.7, 6.5, 12.1) which could account in part for the preferred binding of vanadate to enzymes[17]. In addition, apparently vanadate(V) can readily acquire a stable five-coordinate trigonal bipyramidal geometry resembling the transition state of phosphate as originally proposed by Lindquist and coworkers[26]. The structures of anhydrous and hydrous potassium metavanadate, $KVO_3$ (i) and $KVO_3 \cdot H_2O$ (ii), indicate that both tetrahedral and trigonal bipyramidal structures can be easily achieved[23].

(i)

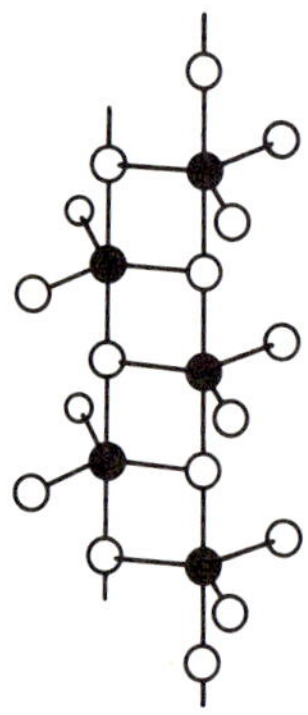

(ii)

Five-coordinate pentavalent vanadium has never been actually demonstrated for any enzyme, nor is the state of protonation of the bound vanadate known. While monomeric orthovanadate, or one of its protonated forms, appears to be the species responsible for enzyme inhibition, in a few instances the decavanadate may be involved[27, 28]. In general, caution must be exercised when attributing inhibition to a particular species of vanadate. Which vanadate species are present in solution depends markedly on pH and on concentration and, moreover, equilibrium is usually only slowly achieved.

Vanadate solutions are usually prepared from sodium orthovanadate, $Na_3VO_4$, or sodium metavanadate, $NaVO_3$, or from the corresponding ammonium salts. These give the same vanadate species in solution after equilibration. In vitro assays of the effect of vanadate reported in the literature have used a wide range of concentrations, ca 10 mM to 1 nM. It is instructive to examine the species of vanadate which are expected to be present under conditions of equilibrium at these concentrations of vanadium.

Figure 2 shows the mole percent distribution of vanadate species present as a function of total vanadium concentration at a fixed pH of 7.4. Because of the monomer-oligomer equilibria present in these solutions, a pronounced change in the distribution occurs in the concentration range $10^{-3}$ to $10^{-5}$ M (pV = 3 to 5). At $10^{-3}$ M, the trimer $V_3O_9^{3-}$ accounts for approximately 85% of the total. At concentrations as low as $10^{-5}$ M, signifi-

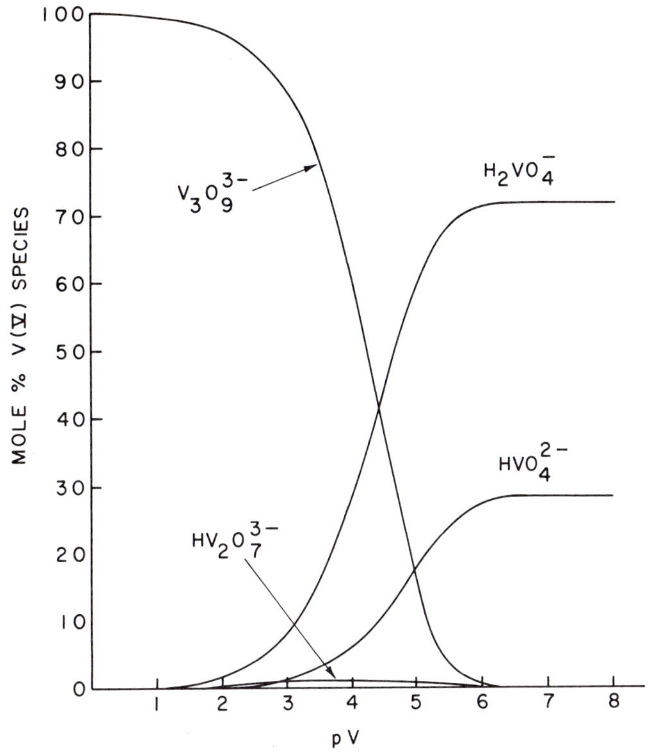


**Fig. 2.** Distribution of vanadate(V) species as a function of total vanadium concentration. Mole percent $V_3O_9^{3-}$ given by $100[V_3O_9^{3-}]/([V_3O_9^{3-}] + [HV_2O_7^{3-}] + [H_2VO_4^-] + [HVO_4^{2-}])$ etc. $pV = -\log[V]_{total}$ where $[V]_{total} = 3[V_3O_9^{3-}] + 2[HV_2O_7]^{3-} + [H_2VO_4^-] + [HVO_4^{2-}]$. *Curves* calculated by the author from equilibrium quotients given in Ref. 24 and 30

cant amounts (~15%) of trimer are still present. Above 0.05 M a tetramer is formed which is not represented in Fig. 2[29].

It is clear from Fig. 2 that oligomeric species are present in studies employing vanadium concentrations greater than nominally $10^{-5}$ M. In the absence of careful concentration and pH dependence measurements, it is not possible in the higher range of concentration to identify the species responsible for activating or inhibiting an enzyme system. Published equilibrium quotients, in principle, should enable one to correct inhibition constants once the inhibiting species is identified[24, 30].

## *II. Vanadium(IV)*

In aqueous solution, the chemistry of tetravalent vanadium is centered around the $VO^{2+}$ ion. This ion forms strong complexes with a diversity of ligands and is known to bind to numerous proteins[21]. The coordination chemistry of $VO^{2+}$, which is relevant to its biochemistry, has been recently reviewed by the author[21]. Five and six coordinate $VO^{2+}$ complexes are formed in which the short vanadyl VO bond length of ~0.16 nm is

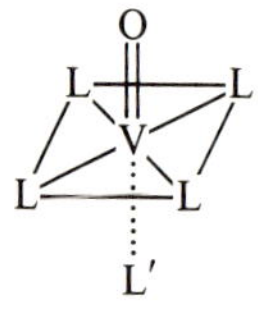
(iii)

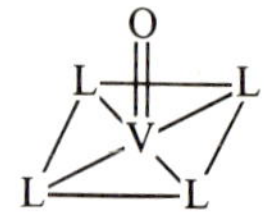
(iv)

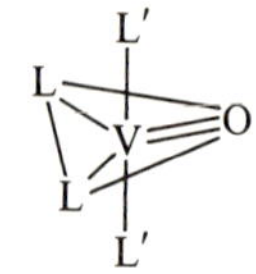
(v)

preserved. Complexes usually approach one of three idealized geometries – square bipyramidal (iii), square pyramidal (iv), or trigonal bipyramidal (v). The latter geometry (v) is much less common for small vanadyl complexes except when dictated by the steric requirements of the ligands themselves. This situation probably prevails to a significant extent in protein complexes. Geometries intermediate between iv and v are also observed. Such flexibility in coordination geometry probably accounts for the fact that $VO^{2+}$ binds to the specific sites of a number of metalloproteins having different metal site structures. The potent inhibition of ribonuclease by a vanadyl(IV)-uridine complex has been ascribed to its ability to adopt a trigonal bipyramidal geometry when bound to the enzyme[26)].

In the absence of strongly complexing ligands, $VO^{2+}$ undergoes a number of hydrolysis reactions. The major hydrolysis products observed in different pH intervals are shown below[21, 30–32)].

| pH 0–3 | pH 3–6 | pH 6–11 | pH 11–14 |
|---|---|---|---|
| $VO^{2+}$ | $VOOH^+$<br>$(VOOH)_2^{2+}$ | $VO(OH)_{2(s)}$<br>+<br>? | $VO(OH)_3^-$ |

The limited solubility of $VO(OH)_{2(s)}$, $K_{sp} \sim 10^{-22}$, can restrict the maximum value of the $VO^{2+}$ concentration which can be used in biological studies (see, for example, Ref. 33). Not all of the hydrolyzed species present in the pH 7–11 interval are known, however. When $VO^{2+}$ is slowly added to Hepes buffer solution or to water maintained at pH ~ 7.5, no precipitate is observed. Attempts in this laboratory to elucidate the soluble species present have been hampered by the slow reversibility of the reactions involved. The species are, however, EPR silent at room temperature and therefore contain more than one metal ion.

The distribution of hydrolyzed $VO^{2+}$ as a function of pH at a total vanadium concentration of 10 μM is shown in Fig. 3. The curves in the distribution diagram also depend on the total vanadium concentration because of the dimer formation and the precipitation reactions. While distribution diagrams of this type for $VO^{2+}$ are incomplete, they nevertheless illustrate the interrelationship between some of the species present and are of predictive value below pH 6 and above pH 11, and possibly in the pH 6 to 11 interval provided one starts with a solution below pH 6 and slowly adds base. The unidentified soluble hydroxide species are less likely to form under those conditions. Species distribution diagrams for a number of $VO^{2+}$ complexes with several common ligands are given by Kraglen[34)].

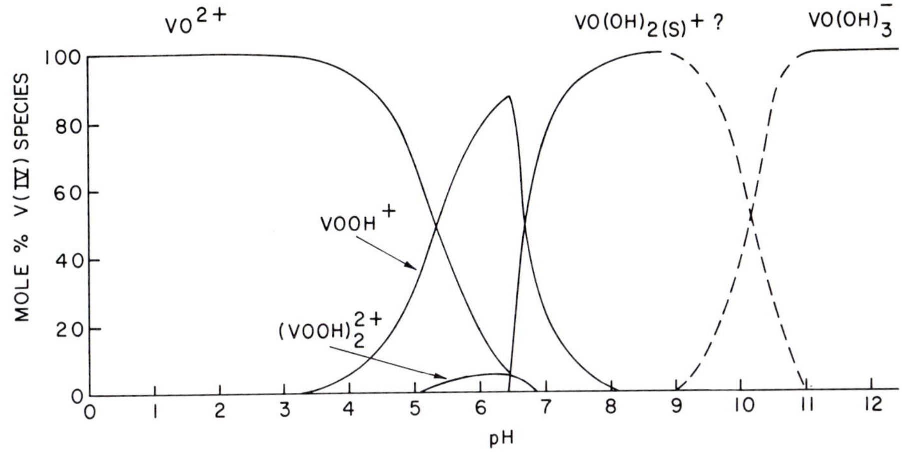


**Fig. 3.** Distribution of vanadyl(IV) species as a function of pH. Quantities defined similarly as in Fig. 2. Here $[V]_{total}$ = 10 μM. Curves calculated by the author from equilibrium quotients reported in reference 21. The $K_{sp}$ for $VO(OH)_{2(s)}$ was taken as $5 \times 10^{-22}$. *Dashed lines* were not calculated. They represent approximate behaviour in pH regions where reliable equilibrium data are not available. An unidentified species indicated by "?" is also present in the pH 7–11 range

Above pH 2–3, solutions of the vanadyl(IV) ion tend to undergo air oxidation to form vanadates (Fig. 1). This susceptibility to oxidation makes biological studies with $VO^{2+}$ more difficult than with vanadate(V) and probably accounts in part for the fact that $VO^{2+}$ has not been extensively studied as an inhibitor. Hydrolyzed species of $VO^{2+}$ are very prone toward air oxidation. When the $VO^{2+}$ ion is chelated, however, oxidation is considerably retarded.

Reduction of vanadate(V) to vanadyl(IV) takes place in the presence of a number of reducing agents common to cells. These include L-ascorbic acid[35)], glutathione[36)], norepinephrine[37)], cysteine[38)], and possibly NADH[39–42)]. The presence of an oxidoreductase enzyme for the vanadium(V)/NADH system has been called into question[40, 42)]. Interestingly, Haight and coworkers have shown that the inorganic redox system $VO^{2+}$ + $H_2O_2$ (or $KMnO_4$) can be coupled to the hydrolysis of ATP[43–45)].

In general, $VO^{2+}$ forms strong complexes with proteins; apparent binding constants are in excess of $10^9$ $M^{-1}$, the minimum value necessary to overcome metal ion hydrolysis at physiological pH ~ 7.5. Because of the high affinity of $VO^{2+}$ for proteins and the fact that it also forms stable complexes with small ligands such as citrate, ATP, $PP_i$, catecholamines, and free amino acids, it is highly unlikely that $VO^{2+}$ exists free in biological systems except perhaps in acidic vacuoles such as the lysosome.

## C. Spectroscopy

### *1. Absorption Spectroscopy*

Vanadium(V), lacking d-electrons, shows no transitions in the visible region of the electromagnetic spectrum. The yellow color observed with some vanadium(V) solutions

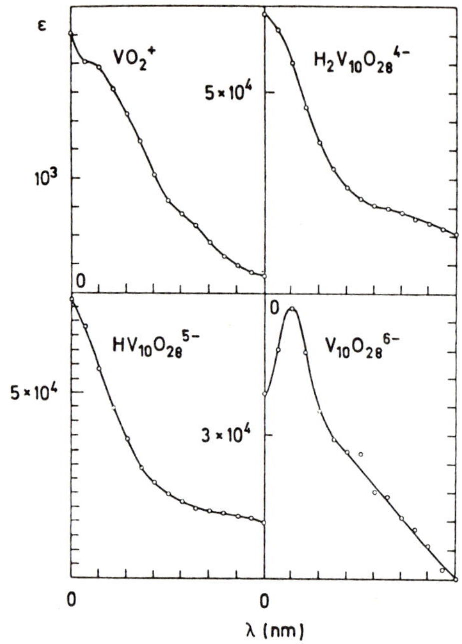


**Fig. 4.** Ultraviolet spectra of species found in acid (pH < 7) vanadate(V) solutions. Wavelength range is 250–390 nm. Reprinted with permission from Ref. 24

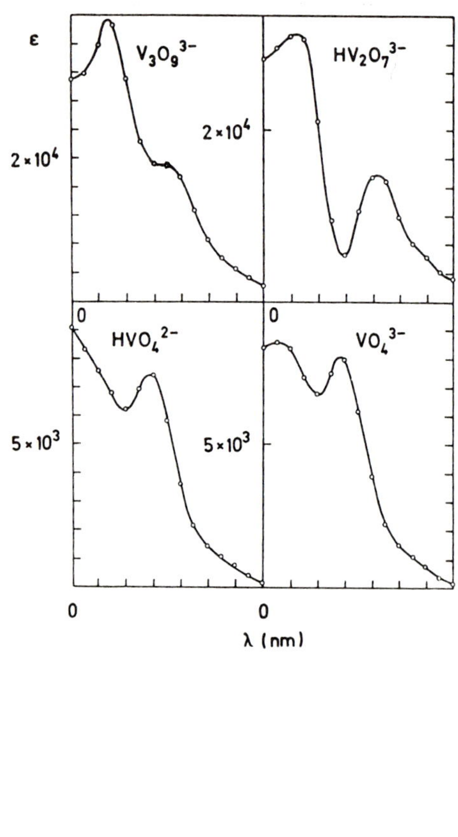


**Fig. 5.** Ultraviolet spectra of species found in basic (pH > 7) vanadate(V) solutions. Wavelength range is 210–350 nm. Reprinted with permission from Ref. 24

is due to an intense absorption tailing in from the ultraviolet. The UV absorption spectra presented in Fig. 4 and 5 of the various vanadate species have been determined by deconvolution techniques[24)]. The spectra, for the most part, are characteristic of individual species and have been used successfully to measure the equilibrium quotients for the formation of the different vanadates[24)]. Spectra of the vanadates would not be easy to measure in the presence of proteins without resorting to difference spectral techniques.

Vanadyl(IV) complexes have a $3d^1$ metal ion electronic configuration which gives rise to ligand field bands in the visible region. Molar absorptivities of typically 150 $M^{-1} cm^{-1}$ or less are observed. The electronic energies generally correlate well with ligand type and with EPR parameters. In a few instances, these spectra have been used to obtain information about the ligand fields of metal sites in proteins[21, 46, 47)]. The optical properties of vanadyl(IV) small chelate and protein complexes have been reviewed[21)].

## *II. NMR Spectroscopy*

The magnetic properties of vanadium are well suited for spectroscopic studies. Vanadium is nearly 100% abundant in the $^{51}V$ (I = 7/2) isotope. Because of the large nuclear moment of $^{51}V$ (5.1392 $\beta_N$), the NMR sensitivity of this nucleus is about 0.38 of that for the proton. Despite the fact that $^{51}V$ has an electric quadrupole moment, relatively narrow lines, ca 60–100 Hz (width at half height) are observed for the tetrahedral or approximately tetrahedral vanadates[48)]. $^{51}V$ spectra have been measured on dilute, ~1 mM, solutions[29)]. NMR studies in solution are limited to the diamagnetic vanadium(V) species, since paramagnetism greatly accelerates nuclear relaxation rates, which results in unobservably broad NMR lines.

As expected for a metal ion possessing a quadrupole moment, complexes of less than perfect cubic symmetry have broadened NMR lines. Thus, the low symmetry $VO_2(EDTA)^{3-}$ complex exhibits a linewidth of 800 Hz which is considerably larger than the 60–200 Hz observed for nearly tetrahedral and octahedral vanadates[48)]. The chemical shift of the vanadates is about 600 ppm. In general, good correlations between vanadium site symmetries (tetrahedral, octahedral, distorted tetrahedral, tetragonal, rhombic), net charge, linewidths, and chemical shifts exist[29, 48)]. However, to date, NMR measurements on a "trigonal bipyramidal" vanadium(V) species have not been made.

## *III. EPR Spectroscopy*

EPR spectroscopy is the usual method chosen to study the $VO^{2+}$ ion and its complexes. The unpaired electron is strongly coupled to the $^{51}V$ nucleus to give EPR spectra consisting of eight line manifolds in frozen solution (Fig. 6). The hyperfine coupling constants and g-factors of the spin Hamiltonian are sensitive to the ligand field of the vanadium. This sensitivity is apparent in the fine structure observed in the EPR spectra of vanadyl transferrin, for example, where three binding environments, A, $B_1$, and $B_2$, are discernable (Fig. 6). The details of the interpretation of spectra such as these are presented elsewhere[21, 49)].

Since EPR spectra of reasonable quality can be obtained on samples containing as little as 20 μM vanadium, studies of $VO^{2+}$ in tissues and subcellular fractions are possi-

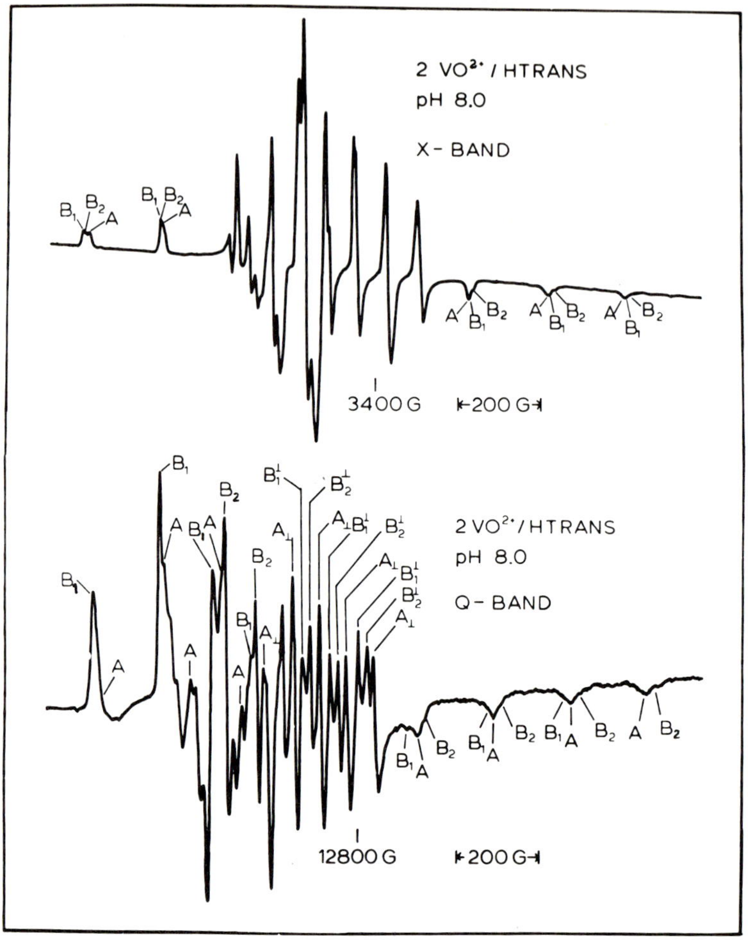


**Fig. 6.** X-band (9.2 GHz) and Q-band (35 GHz) EPR spectra of frozen solutions of divanadyl(IV) human serum transferrin at 77 K and 110 K respectively. A, $B_1$, and $B_2$ designate features in the spectra attributable to three different binding environments of the metal. Reproduced with permission from Ref. 49

ble. Investigations of this type have been carried out on rat liver cells, and on mitochondrial, microsomal supernatant, and microsomal fractions thereof[50, 51]. These experiments have shown that intraperitoneal injection of vanadium(V) results in deposition of vanadium(IV) in the liver as $VO^{2+}$. The EPR parameters, $g_0$ and the hyperfine coupling constant $A_0$ of $^{51}V$, derived from the relationships $g_0 = g_\parallel/3 + 2g_\perp/3$ and $A_0 = A_\parallel/3 + 2A_\perp/3$, suggest that most of the donor atoms are oxygen. Correlation diagrams of EPR parameters with ligand type are given in a number of places[21, 49, 51–53]. It should be noted that a common mistake in the calculation of $A_0$ from the above relationship is to express $A_\parallel$ and $A_\perp$ in units of gauss rather than in units of energy (Hz or $cm^{-1}$). Because of the anisotropy in $g$, the result can be significantly in error.

EPR spectra of $VO^{2+}$ in room-temperature solutions are observable because the non-degenerate ground state of the unpaired electron has no excited states nearby in energy.

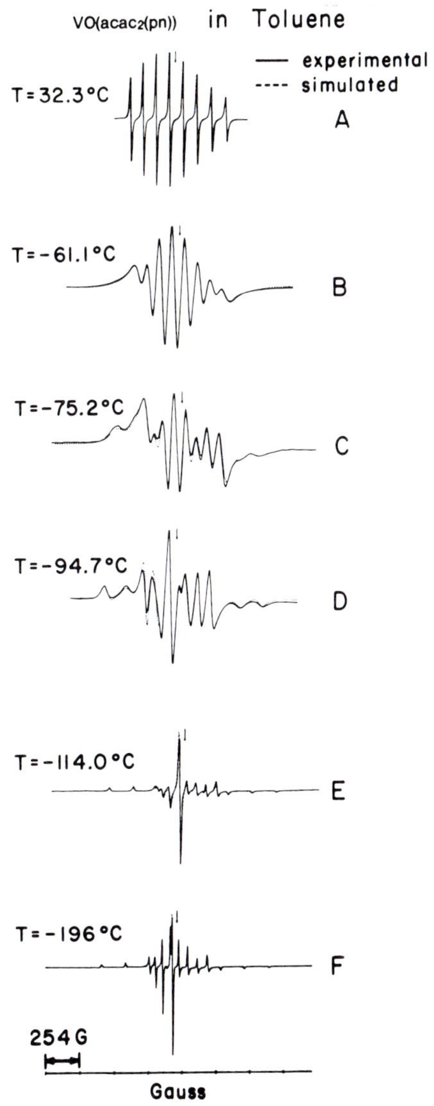


**Fig. 7 A–F.** Comparison of experimental and simulated spectra from the rapid motional to the rigid limit for $VO(acac_2(pn))$ in toluene. All simulations use a Brownian rotational diffusion model with $\tau_R^B$ defining the rotational correlation time. $\alpha''$ is the residual line-width contribution. **(A)** $\tau_R^B = 2.06 \times 10^{-11}$ s, $\alpha'' = 1.7$ G; **(B)** $2.63 \times 10^{-10}$ s, 1.7 G; **(C)** $5.00 \times 10^{-10}$ s, 1.7 G; **(D)** $2.25 \times 10^{-10}$ s, 1.7 G; **(E)** $5.0 \times 10^{-8}$ s, $\alpha'' = 6$ G; **(F)** rigid limit. Reproduced with permission from Ref. 54

The EPR spectrum, through the rotational correlation time, is sensitive to the size of the complex and the solution viscosity. Vanadyl protein complexes in solution at room-temperature display near rigid limit spectra due to slow molecular tumbling. Figure 7 shows the effect of viscosity changes in the EPR spectrum of bisacetylacetonepropylenediimineoxovanadium(IV), $VO(acac_2(pn))$. Spectra ranging from the motionally narrow limit (A) to the rigid limit (F) are obtained[54)]. The lineshape of spectrum E is typical of that obtained with proteins, where the rotational correlation time of the metal ion is governed by the rotational diffusion of the macromolecule[55–58)].

## IV. ENDOR Spectroscopy and ESE

The ground state of $VO^{2+}$ complexes is essentially a metal based nonbonding $d_{xy}$ orbital. Because the unpaired electron only weakly interacts with the ligand nuclei, the observation of superhyperfine interactions in the EPR spectrum is relatively rare. In most instances the magnitude of the coupling is less than the natural EPR linewidth. This difficulty has been overcome by van Willigen and coworkers through the use of ENDOR[59–62]. They have nicely shown that anisotropic ligand superhyperfine coupling constants of $^{14}N$ and $^{1}H$ as well as $^{14}N$ quadrupole coupling constants can be extracted from the ENDOR spectrum. The VO bond axis defines a unique "parallel" direction in all vanadyl complexes, a direction which can be identified in the EPR spectrum of randomly oriented molecules. Thus, in the ENDOR experiment, one can selectively irradiate molecules with their vanadyl VO bonds oriented parallel or perpendicular to the direction of the static magnetic field. The $^{14}N$ and pyrrole $^{1}H$ ENDOR spectra of selectively irradiated vanadyl tetraphenylporphyrin are given in Figs. 8 and 9. The superhyperfine coupling constants along these directions have been measured. Through point dipole calculations, one can obtain approximate distance and orientation information about the ligand nuclei relative to the metal center. Similar information has been obtained using the electron spin echo (ESE) technique[63]. The application of ENDOR and ESE spectroscopies to the study of the metal site structures of $VO^{2+}$ protein complexes is obvious. ESE of metalloproteins has been recently reviewed[64].

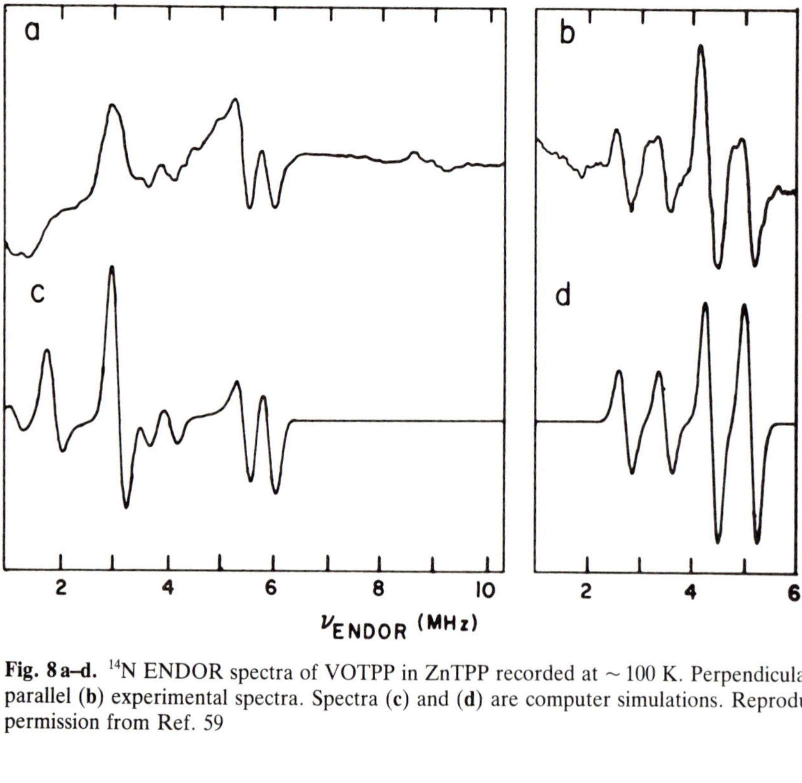


**Fig. 8a–d.** $^{14}N$ ENDOR spectra of VOTPP in ZnTPP recorded at ~ 100 K. Perpendicular (**a**) and parallel (**b**) experimental spectra. Spectra (**c**) and (**d**) are computer simulations. Reproduced with permission from Ref. 59

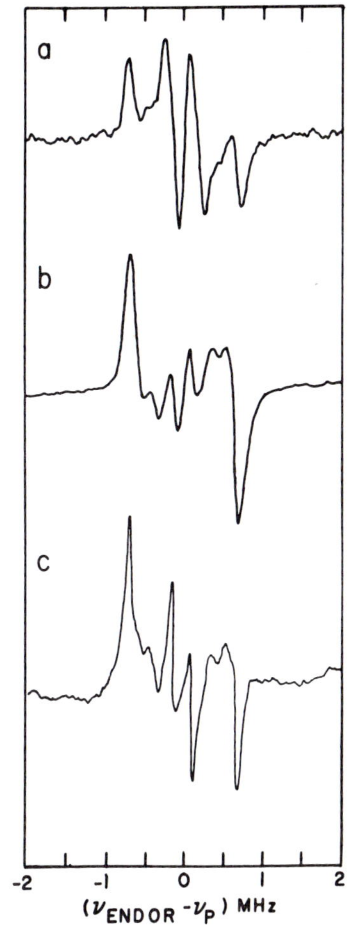


**Fig. 9 a–c.** Perpendicular $^1$H ENDOR spectra of VOTPP in $CDCl_3$-toluene (perdeuterated): (**a**) normal VOTPP at ~ 100 K; (**b**) VOTPP with deuterated phenyl groups at ~ 100 K; (**c**) VOTPP with deuterated phenyl groups at ~ 15 K. Reproduced with permission from Ref. 59

## *V. X-ray Absorption Spectroscopy*

Extended X-ray absorption fine structure (EXAFS) and K edge absorption spectroscopy have been used to study vanadium in coal[7)]. In contrast to the common belief that vanadyl porphyrins are the major compounds present, these studies demonstrate that in the high vanadium content samples examined, the metal is tetravalent and coordinated to oxygen ligands. Similar investigations have been carried out on the vanadocytes of ascidians to identify the form of vanadium in tunicate blood cells[22)] (vide infra).

# D. Biochemical and Physiological Effects

## *I. Nucleases and Phosphatases*

The inhibition of ribonuclease by vanadium was the first known example of the influence of this metal on the activity of an enzyme[26)]. The constants for the competitive inhibition by the 1 : 1 complexes of $VO^{2+}$ or vanadium(V) with uridine are given in Table 2. These inhibitors bind at the active site three orders of magnitude more tightly than does the substrate. In addition to this classic example of nuclease inhibition by vanadium, the complexes of $VO^{2+}$ with thymidine and thymidine-3'-phosphate are even more potent competitive inhibitors of the enzyme staphylococcal nuclease (Table 2)[65, 66]. In this instance the EPR spectra of the inhibitor and the inhibitor-enzyme complex have been recorded[65]. The suppressant effect of vanadyl(IV) ribonucleoside complexes on ribonuclease activity of cell homogenates has been very useful in isolating intact mRNA from lymphocytes[67]. Traditional RNase inhibitors had been found to be ineffective for this work. A vanadyl(IV) ribonucleoside complex is now commercially available as an RNase inhibitor (Bethesda Research Laboratories, Inc.).

**Table 2.** Vanadium inhibition constants for various enzymes

| Enzyme | Form of vanadium[a] | $K_I(M)$[b] | Reference |
|---|---|---|---|
| adenylate cyclase | vanadate(V) | activation | 130 |
| adenylate kinase (skeletal muscle) | decavanadate(V) | $<1 \times 10^{-6}$ | 28 |
| acid phosphatase (human liver) | vanadate(V) | $2 \times 10^{-7}$ | 70 |
| acid phosphatase (wheat germ) | vandate(V) | $6.7 \times 10^{-6}$ | 70 |
| alkaline phosphatase (*E. coli*) | vanadate(V) | $4 \times 10^{-7}$ | 68 |
| alkaline phosphatase (human kidney, liver, intestine) | vanadate(V) | $(0.5–0.9) \times 10^{-6}$ | 69 |
| ATPase (mitochondria) | vanadate(V) | $\sim 1.5 \times 10^{-3}$ | 133 |
| $Ca^{2+}$, $Mg^{2+}$-ATPase (cardiac sarcoplasmic reticulum) | vanadate(V) | $(5–50) \times 10^{-6}$ | 91, 133 |
| $Ca^{2+}$ uptake (sarcoplasmic reticulum) | vanadate(V) | $\sim 2 \times 10^{-5}$ | 133 |
| $Ca^{2+}$, $Mg^{2+}$-ATPase (red cell) | vanadate(V) | $0.8 \times 10^{-6}$, $1.5 \times 10^{-6}$ | 87, 88, 90 |
| $Ca^{2+}$-ATPase (squid axon) | vanadate(V) | $7 \times 10^{-6}$ | 101 |
| dynein ATPase (sea urchin sperm) | vanadate(V) | $\sim 10^{-6}$ | 96 |
| dynein ATPase (*Tetrahymena*) | vanadate(V) | $1 \times 10^{-8}–5 \times 10^{-6}$ | 97, 98 |
| $Mg^{2+}$-ATPase (yeast plasma membrane) | vanadate(V) | $(0.14–0.20) \times 10^{-6}$, $1 \times 10^{-6}$ | 93 |
| $Mg^{2+}$-ATPase (Neurospora crassa) | vanadate(V) | $(0.45–1.0) \times 10^{-6}$ | 92 |

**Table 2** (continued)

| Enzyme | Form of vanadium[a] | $K_I(M)^b$ | Reference |
|---|---|---|---|
| myosin ATPase (rabbit muscle) | vanadate(V) | $\leqslant 10^{-6}$ | 95, 199 |
| $Na^+$,$K^+$-ATPase (kidney) | vanadyl(IV) | $8 \times 10^{-6}$ | 133 |
| $Na^+$,$K^+$-ATPase (heart atria and ventricle) | vanadate(V) | $0.6 \times 10^{-6}$ | 148 |
| $Na^+$,$K^+$-ATPase (skeletal muscle membrane) | vanadate(V) | $1 \times 10^{-6}$ | 76 |
| $Na^+$, $K^+$-ATPase (sarcolemma) | vanadate(V) | $1 \times 10^{-6}$ | 200 |
| $K^+$-p-nitrophenol phosphatase (skeletal muscle membrane) | vanadyl(IV) and vanadate(V) | $4 \times 10^{-8}$[c] | 76 |
| phosphofructokinase | decavanadate(V) | $4.5 \times 10^{-8}$, $5.5 \times 10^{-7}$ | 27 |
| phosphoglycerate mutase (BPG dependent) | vanadate(V) | $5 \times 10^{-6}$ | 121 |
| phosphoglucomutase | vanadate(V) | 200-fold enhancement | 122 |
| ribonuclease | vanadyl(IV) | $6 \times 10^{-5}$ | 26 |
| ribonuclease | vanadyl(IV) : uridine | $1 \times 10^{-5}$ | 26 |
| ribonuclease | vanadium(V) : uridine | $1.2 \times 10^{-5}$ | 26 |
| staphylococcal nuclease | $Ca^{2+}/VO^{2+}$/thymidine | $8 \times 10^{-8}$ | 65, 66 |
| staphylococcal nuclease | $Ca^{2+}/VO^{2+}$/thymidine-3-phosphate | $1.7 \times 10^{-8}$ | 65, 66 |

[a] In most instances the chemical identity of the inhibitor is not known with certainty. It is generally believed that most vanadate(V) inhibitors correspond to one of the protonated forms of ortho-vanadate(V)

[b] Inhibitor dissociation constant. Some values listed correspond in the original work to the concentration of inhibitor which reduced the observed enzyme catalyzed rate by 50%. In these cases, kinetic analyses using the standard equations for enzyme inhibition were not employed

[c] Same inhibitor constant reported for both

Acid and alkaline phosphatases which are believed to pass through a phosphorylated intermediate during the hydrolysis of p-nitrophenolphosphate are inhibited by vanadate(V)[68–69]. In the case of alkaline phosphatase from *E. Coli,* $VO^{2+}$ inhibits better than vanadate(V) (Table 2)[68]. Typical inhibitor plots are shown in Fig. 10. The reduction of the effectiveness of $VO^{2+}$ as an inhibitor by about a factor of five in tris buffer probably stems from the complexation of the metal ion by tris. Such complexation is known to occur[21].

VanEtten and coworkers[70] have conducted a comparative study of the relative effectiveness of various phosphate analogues as inhibitors of acid phosphatases. Their results are summarized in Table 3. All of these inhibitor ions are isostructural with the product

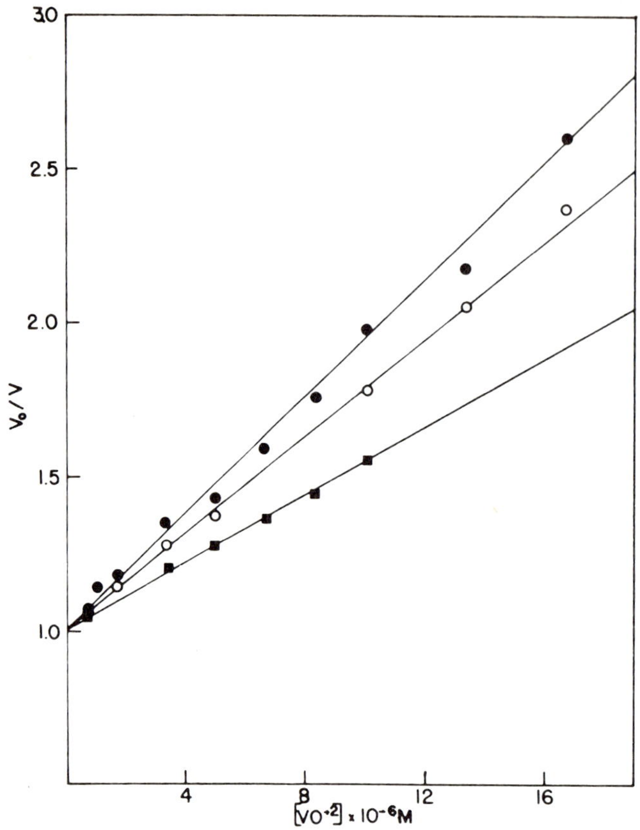


**Fig. 10.** Inhibition of alkaline phosphatase by $VO^{2+}$ in 0.02 M Tris, 0.5 M KCl, pH 8.0, PNPP, $2 \times 10^{-5}$ M (●), and $5 \times 10^{-5}$ M (■) and in 0.02 M ethanolamine, 0.5 M KCl, pH 10.0, PNPP, $5 \times 10^{-5}$ M (○). Reproduced with permission from Ref. 68

**Table 3.** Constants for competitive inhibition of acid phosphatases by phosphate and its analogues[a]

| Inhibitor | $K_I(M)$[b] Human liver | Wheat germ |
|---|---|---|
| orthophosphate | $8.0 \times 10^{-4}$ | $1.3 \times 10^{-3}$ |
| arsenate | $1.8 \times 10^{-4}$ | $6.1 \times 10^{-4}$ |
| vanadate(V) | $2.0 \times 10^{-7}$ | $6.7 \times 10^{-6}$ |
| molybdate(VI) | $3.6 \times 10^{-8}$ | $5.6 \times 10^{-8}$ |
| tungstate | $1.6 \times 10^{-8}$ | $6.1 \times 10^{-8}$ |

[a] Data from VanEtten and coworkers[70)]
[b] Inhibition constant for dissociation

inhibitor orthophosphate. Bond lengths increase progressively with the $K_I$ value and range from 0.152 nm for P–O of phosphate to 0.176 nm for W–O of tungstate. The authors argue that variations in bond length alone cannot account for the observed differences in $K_I$ values. They suggest that the ability of the inhibitor to coordinate to the enzyme and adopt a trigonal bipyramidal geometry is also important.

## *II. ATPases*

### *1. $Na^+$, $K^+$-ATPase: The Sodium Pump*

The early work on the strong inhibitory effect of vanadium on nucleases and phosphatases suggested that vanadate(V) and vanadyl(IV) ions might be useful as kinetic probes of the mechanisms of other enzymatic reactions involving the hydrolysis of phosphate ester bonds. However, these studies received relatively little attention until 1977. At that time, Cantley and coworkers[71)] identified an impurity in commercial ATP (Sigma) prepared from equine muscle as orthovanadate(V), a potent inhibitor of $Na^+$, $K^+$-ATPase. For over a decade, other workers had observed an inhibitory effect due to this elusive impurity in ATP isolated from biological sources but not in laboratory synthesized ATP. It is now known that orthovanadate(V) copurifies with Sigma grade ATP at a V : ATP molar ratio of about 1 : 3000. This naturally occurring inhibitor has been used extensively to study the mechanism of the sodium pump.

The intracellular concentration of $K^+$ is high in most animal cells relative to that of external fluids. The converse is true for $Na^+$. The free energy necessary to pump these ions against their concentration gradients is derived from the hydrolysis of ATP. Hydrolysis is controlled by the $K^+$ and $Na^+$ stimulated ATPase which is imbedded in the cell membrane and has stimulation sites on both the intracellular and extracellular surfaces. To fully activate the pump, $Mg^{2+}$, in addition to $Na^+$ and $K^+$, must be present. $K^+$ activates the ATPase exterior to the cell and $Na^+$ on the interior. An introduction to this subject can be found in Stryer[72)]. The effect of vanadate on $Na^+$, $K^+$-ATPase has been reviewed[19, 73)].

The overall reaction consists of phosphorylation and dephosphorylation steps.

$$E + ATP \xrightleftharpoons{Na^+, Mg^{++}} E{-}P + ADP$$

$$E{-}P + H_2O \xrightarrow{K^+} E + P$$

$Na^+$ and $Mg^{++}$ stimulate the enzyme phosphorylation reaction generating phosphorylated enzyme intermediate E–P (first step) which is dephosphorylated in the presence of $K^+$ (second step). The net result is the transport of $3\,Na^+$ out of the cell and $2\,K^+$ ions into the cell per ATP molecule hydrolyzed. ATP binds and is hydrolyzed on the cytoplasmic side of the membrane. A simplified mechanism for the overall process is shown below where $E_1$ and $E_2$ represent two conformations of the protein which have high affinities for $Na^+$ and $K^+$ respectively. Complete details of how the ATPase/pump system operates are lacking, but probe experiments using orthovanadate(V) have provided some insight into the mechanism.

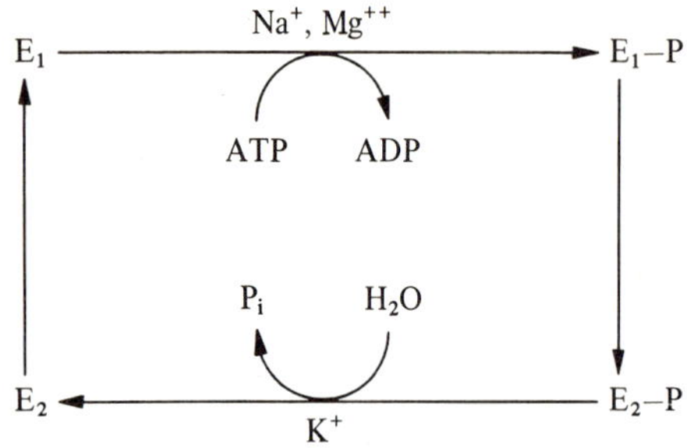


Orthovanadate(V) enters the red cell via the same anion transport system used by orthophosphate[74)]. The process of achieving equal concentrations inside and out has a half life of only 4 min followed by a slower phase $t_{1/2} \sim 90$ min in which the cell continues to accumulate vanadium, finally reaching a concentration gradient of 500 μM/40 μM inside/outside. This internalized vanadate(V) has been shown by using EPR to be slowly reduced to $VO^{2+}$ by glutathione, which then binds quantitatively to hemoglobin [36, 58)]. The reduction and complexation of the resultant $VO^{2+}$ is evidently the driving force for accumulation of vanadium by the cell. The site(s) of binding to the protein are not known, nor is it known whether the association of $VO^{2+}$ with hemoglobin has any biochemical significance.

The $Na^+$, $K^+$-ATPase is inhibited by internalized vanadate from the cytoplasmic side of the membrane[74)]. This inhibition is controlled and facilitated by $K^+$ binding at an extracellular site[75)]. Norepinephrine, by complexing and reducing vanadate(V) to vanadyl(IV), reverses the inhibition[37)]. The vanadyl(IV) ion is not a good inhibitor of the enzyme, being at least an order of magnitude less effective than $VO_4^{3-}$. In contrast it has been recently reported that $VO^{2+}$ inhibits p-nitrophenolphosphatase activity of diaphragm membrane equally well as $VO_4^{3-}$ [76)].

A number of mechanistic studies using vanadate as a probe of $Na^+$, $K^+$-ATPase have appeared (see, for example, Refs. 74–82). Vanadate and the γ phosphate of ATP both bind at a common site on the protein, apparently forming a covalent link with aspartate[78, 81)]. A divalent cation, $Mg^{2+}$ or $Mn^{2+}$, binds synergistically to the protein along with vanadate and is not released until vanadate dissociates[78, 82)]. ATP does not bind to the vanadate trapped enzyme and cannot phosphorylate it (first step, above). In forming the putative stable trigonal bipyramidal intermediate vanadate enzyme complex, the enzyme appears to be locked into the $E_2$ conformation[78, 79)]. When phosphate is bound to the enzyme instead of vanadate, this conformation leads to dephosphorylation (second step, above).

The vanadate inhibition is facilitated by $K^+$ and $Mg^{2+}$ as noted earlier, but there is conflicting evidence as to whether the inhibitory sites of $K^+$ and $Mg^{2+}$ are distinct from the sites which activate the enzyme in the absence of vanadate[77, 82)]. Ouabain, a cardiac glycoside, has a well known inhibitory effect on $Na^+$, $K^+$-ATPase. The binding of this agent at an extracellular site of the enzyme is modulated by $K^+$ as is the binding of vanadate at the intracellular site. The $K^+$ site may be the same in both instances[77)]. High concentrations of sodium reverses vanadate inhibition. Cantley and coworkers[78)], by taking into account a number of observations in their laboratory and elsewhere, have proposed a detailed model of the catalytic cycle of $Na^+$, $K^+$-ATPase showing explicitly

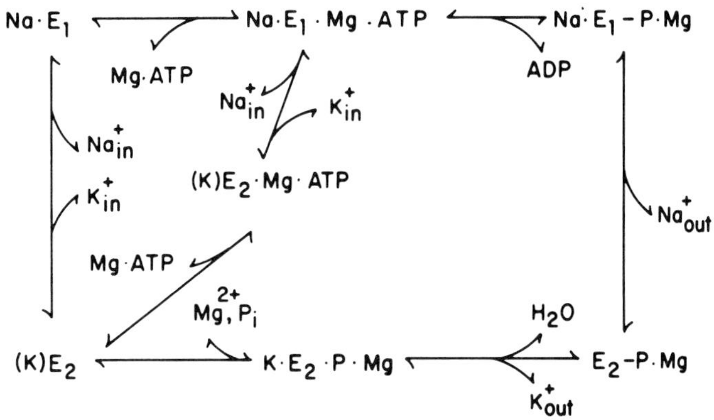


**Fig. 11.** A mechanism for the (Na,K)-ATPase explaining different rate-limiting steps at high and low ATP concentrations. At low ATP concentrations, $Na^+$ transport out of the cell, $K^+$ transport into the cell, and ATP hydrolysis occur as indicated by a clockwise direction through the outer cycle. At high ATP concentrations, ATP binds to (K)$E_2$ and the inner cycle is predominant. *Dashes* indicate covalent bonds and *dots* indicate noncovalent bonds. Vanadate is proposed to bind in place of phosphate at the K · $E_2$ · P · Mg step. Reproduced with permission from Ref. 78

the binding of substrates, inhibitors and metal ions (Fig. 11). During inhibition, the intermediate K · $E_2$ · P · Mg in Fig. 11 is replaced by K · $E_2$ · V · Mg which stabilizes the $E_2$ conformation, effectively blocking the cyclic operation of the pump. This intermediate complex is slow to dissociate. A somewhat similar mechanism has been suggested by Karlish et al.[83].

The observation that vanadium is present in most tissues led to the suggestion a few years ago that the metal may have a role in regulation of the sodium pump in vivo[71]. Vanadate(V) is a potent inhibitor, requiring only minute concentrations to be effective (Table 2), and inhibition can be essentially reversed through reduction of vanadate(V) to vanadyl(IV). Subsequently, it has been shown that numerous other enzymes are also inhibited by vanadate (Table 2). In light of these results, vanadate does not exhibit the specificity that one would prefer for a regulator. In addition, the intracellular form of vanadium appears to be largely vanadyl(IV)[36]. Although function is difficult to prove, one view is that vanadium has several biological roles and is inhibitory for many enzymes. Regardless of whether endogenous vanadate is ultimately proven to be important physiologically, this anion has been of immense value in probing complex enzymatic processes.

The inhibition of $Na^+$, $K^+$-ATPase by $VO_4^{3-}$ has been developed as an assay for vanadium in serum samples[84, 85].

## 2. *Other ATPases*

The results of the studies with $Na^+$, $K^+$-ATPases prompted workers to examine other ATPases for effects of vanadate(V). $Ca^{2+}$-ATPase of red cell membranes, which is coupled to the calcium pump, is also strongly inhibited by vanadate[86–88], but early work suggested that $Ca^{2+}$-ATPase from the sarcoplasmic reticulum of muscle is not[89]. The result of this latter study is at variance with more recent work[90, 91]. ATPase of yeast plasma membrane and *Neurospora crassa* membrane are inhibited by vanadate and like the $Na^+$, $K^+$-ATPase, $Mg^{2+}$ binding is required[92–94]. Myosin[95, 199] and dynein ATPases of muscle, and of flagella and cilia respectively are also inhibited[96–99] as is the ATP promoted $Na_0^+$-dependent $Ca^{2+}$ efflux from barnacle muscle fibers[100] and $Ca^{2+}$ efflux in squid axons[101]. Earlier work suggested that myosin ATPase was not affected[89]. $F_1$-ATPase and actomyosin ATPase from mitochondria does not appear to be inhibited by vanadate nor does $Ca^{2+}$-ATPase from ascites plasma membrane[89, 90, 96].

## *III. Insulin-Like Activity of Vanadium*

Several laboratories have recently shown that vanadium is an insulin mimetic agent[33, 102, 103]. The mechanism of action of insulin in the cellular uptake and metabolism of glucose and in other transport processes is not yet known. Substances which mimic or enhance the action of insulin are of interest because of their value in elucidating the molecular origin of hormone action. Because micromolar concentrations of vanadium are found in cells having insulin receptors, the possibility of an in-vivo role is again raised.

The incubation of rat adipocytes, one of the prime target cells of insulin, with 10–100 μM concentrations of vanadyl(IV) or vanadate(V) stimulates the transport of glucose and 2-deoxyglucose into the cell[33, 102]. The maximal stimulation of hexose transport due to vanadium alone is essentially equivalent to the maximal stimulation produced by insulin in the absence of vanadium. Glucose oxidation is also stimulated. The insulin-like activity correlates with the vanadium content of the cells, but the activity of $Na^+$, $K^+$-ATPase does not[102]. This result suggests that the active form of the vanadium in these experiments may not be vanadate(V) but rather vanadyl(IV). It is difficult, however, to unequivocally establish the oxidation state and chemical form of the vanadium responsible for the effect. The suggestion that vanadyl(IV) is operating here is supported by the work of Schecter and Karlish who found that glucose oxidation is stimulated more in adipocytes incubated with vanadate(V) plus glutathione or epinephrine, reductants for vanadium(V), than with vanadate(V) alone[33]. GSH has no effect when the cells are incubated with vanadyl(IV). Vanadate(V) has been shown to be incorporated into the adipocyte and converted to two different EPR observable $VO^{2+}$-GSH complexes[206].

It is interesting that externally applied $VO^{2+}$ is less effective than $HVO_4^{2-}$ in causing an insulin-like response in adipocytes[33]. It may be that $VO_4^{3-}$ penetrates the cells rapidly via the anion transport system, and once inside is reduced to $VO^{2+}$, which then causes the insulin response. Alternatively, much of the externally added $VO^{2+}$ might be bound to the albumin of the Krebs-Ringer solution in which the cells are suspended, making less $VO^{2+}$ available for cellular uptake. $VO^{2+}$ is known to bind stoichiometrically to serum albumin[55].

The mode of action of vanadyl(IV) ion in mimicking insulin is not clearly defined. It has been speculated that $VO^{2+}$ may be inhibiting cellular phosphatases[33]. Since alkaline phosphatase is strongly inhibited by $VO^{2+}$ [68], the metal may have a functional role in regulating phosphatase activity in cells. In a very recent development, insulin has been found to stimulate phosphorylation with orthophosphate of the 95,000 molecular weight β subunit of the insulin receptor[198]. The phosphorylation of the receptor could be an early step in the mechanism of insulin action and the step where vanadium intervenes.

It has been reported that vanadate(V) enhances the stimulatory effects of insulin on DNA synthesis in cultured mammary gland cells of mice[103]. Vanadate(V) in the absence of insulin has little effect. Insulin and vanadate(V) act as comitogens. The reported experiments do not preclude the possibility that vanadyl(IV) is the form of vanadium responsible for these effects. Vanadate(V) has also been reported to be an epidermal growth factor stimulating DNA synthesis[104].

Like insulin, vanadate inhibits protein degradation in hepatocytes[105]. Proteolysis in liver homogenates at pH 5 is inhibited, suggesting that the lysosomal proteinases are affected.

## *IV. Accumulators of Vanadium*

Few organisms concentrate vanadium to any appreciable extent. Well known exceptions are the mushroom *A. muscaria*[106] and the ascidians[20]. An unusual vanadyl(IV) complex has been isolated from the cap of *A. muscaria* and characterized by EPR and chemical methods[107]; however, it has been suggested by the original investigators and others that this complex may not correspond to the original substance present[17, 107]. The limited information available on vanadium in *Amanita* is discussed by Lancashire[108].

The ascidians, or tunicates (sea squirts), perform the remarkable task of concentrating vanadium several million fold over the nominally 50 nM level in seawater[20]. Vanadium as $HVO_4^{2-}$ or $H_2VO_4^-$ is assimilated from seawater pumped through the animal[109]. The metal is accumulated in a reduced state at concentrations of 1 mM to 1 M, depending on the species, in blood cells called vanadocytes[109–112]. NMR, EPR, and EXAFS measurements on whole vanadocyte cells of *Ascidia ceratodes* and *Ascidia nigra* demonstrate that the metal exists mainly as aquated V(III) probably complexed with sulfate with some V(IV) also present[22, 110–113]. In *Ciona intestinalis,* however, the metal appears to be mostly vanadium(IV)[111].

A chromogen, tunichrome, has been isolated from the cells of several species and has been partially characterized[114, 115]. In *A. nigra,* tunichrome is present in near equimolar amounts with vanadium and is capable of reducing vanadate(V). The precise role of tunichrome is not known. It may participate in the accumulation and reduction of vanadium by the cell. A model for cellular uptake of vanadium has been proposed[113, 114].

The function of the vanadocyte remains a mystery. Much of the earlier literature refers to the presence in these cells of a vanadium protein complex hemovanadin or a small organic complex which has an oxygen transport function analogous to hemoglobin. It is now known that such complexes do not exist (vide supra)[116]. Furthermore, it has been clearly demonstrated that vanadocytes of *A. nigra* do not transport oxygen[117]. Perhaps their physiological function is related to the storage and transport of reducing equivalents as V(III). It has been suggested that the vanadium content of ascidians serves as a feeding deterent to predators[118] or may be involved in the synthesis of the tunic[16].

The early literature on tunicates has been reviewed by Biggs and Swinehart[20]. A current review is forthcoming in this series (Kustin and McLeod, in preparation).

## V. Other Biochemical and Physiological Processes

There are a few examples of enzyme inhibition by vanadium via an allosteric effect. Orthovanadate(V) is ineffective at inhibiting phosphofructosekinase, whereas the decavanadate strongly inhibits the enzyme; in fact decavanadate is the most potent inhibitor known[27, 119)]. ATP and decavanadate, both allosteric inhibitors, act synergistically. The inhibition is antagonized by positive effectors such as cAMP. In this instance, the action of vanadate does not reside in the adoption of the putative trigonal bipyramidal geometry about the vanadium upon binding to the enzyme. The polyvalent anionic structure of decavanadate is evidently the important feature of this substance for inhibition. The fact that other polyvalent agents such as polyphosphates and polyadenylic acid are also inhibitory is consistent with this interpretation[27)].

Decavanadate also inhibits adenylate kinase which catalyzes the formation of two moles of ADP from ATP plus AMP[20)]. The binding site for the polynegative decavanadate (or an oligomer of smaller size) on this enzyme has been located by X-ray crystallography in a deep cleft lined with positively charged side chains of a lysine and six arginines[120)]. Since this is the same site in which the two substrates bind, the origin of the inhibition by decavanadate is clear.

It is interesting that vanadate(V) inhibits the 2,3-diphosphoglycerate dependent phosphoglycerate mutases but not the independent phosphoglycerate mutases. It is therefore possible to easily differentiate between the two types of enzymes[121)]. The transfer reaction in phosphoglucomutase, however, is augmented 200-fold by 1 mM vanadate, apparently by lowering the $pK_a$ of tyrosine at the active site[122)]. In general, the enzymes involved in glucose metabolism which catalyze phosphoryl transfer reactions, mutases, kinases, and phosphatases, can be probed by orthovanadate(V)[119)].

Vanadate(V) increases cyclic AMP levels and protein kinase activity in liver slices[123)]. It should be noted again, however, that in experiments such as these involving whole cells, the active form of the vanadium is open to question.

The natriuresis and diuresis effect of vanadate, which concentrates in the proximal tubules of the kidney, is believed to be due to inhibition of the $Na^+$, $K^+$-ATPase of these tissues[19, 124, 125)]. In contrast, the oubain insensitive vanadate inhibition of $HCO_3^-$, $Cl^-$, $Na^+$, and $H^+$ transport in bladder occurs by another, yet unknown, mechanism[126–128)]. Vanadate effects on salt and water effluxes in gills have also been studied[129)].

Contrary to the usual situation of enzyme inhibition, vanadate(V) is found to activate the membrane bound enzyme adenylate cyclase, which catalyzes the formation of cAMP[130)]. This finding may be quite significant owing to the importance of cAMP levels in cellular metabolism. Vanadate appears to activate the enzyme by binding at the GTP site. Currently, a variety of tissues are being extensively studied, most notably those which have a cardiovascular function. Both positive and negative ionotropic effects (strength of contraction) of vanadate on various types of cardiovascular muscle have been noted[39, 130–152)]. Vanadate effects on adenylate cyclase and ATPases have been implicated in these processes.

Vanadate(V) and GTP-γ-S, a hydrolysis resistant analog of GTP, have recently been shown to excite photoreceptors of the horseshoe crab *Limulus* in the dark and have been used to help elucidate the chemical steps which are involved in the phototransduction process[153)]. The results suggest that vanadate binds to a guanyl nucleotide binding protein which activates adenylate cyclase.

Evidence for the involvement of vanadium in the metabolism of molybdenum in *Aztoabacter* was discovered several years ago[154, 155]. Vanadium increases nitrogenase activity of the organism through enhancement of molybdenum utilization under low molybdenum conditions. A V-nitrogenase in which vanadium was substituted for molybdenum was isolated. In another case, vanadium inhibition of nitrate reductase of *Neurospora* was observed[156]. More recently, vanadate has been shown to inhibit nitrate reductase and xanthine oxidase[157, 158].

Lowered levels of cholesterol, triglycerides, and phospholipid biosynthesis due to vanadium were discovered several years ago (see Refs. 3, 4, and 14). It has recently been found that vanadate(V) at concentrations of 1 mM inhibits mevalonate synthesis, the rate limiting step in cholesterol biosynthesis[159]. In this instance the active form of the metal may be vanadyl(IV).

Positive correlations between plasma levels of vanadium and manic depressive psychosis have been observed in a number of laboratories[160–163]. Symptoms are augmented when patients are administered vanadium for 10 days[162], but large doses of ascorbic acid reverses this effect possibly through the reduction of vanadium(V) to vanadium(IV)[160]. It has been suggested that manic depressives may have a defect in the sodium pump or that vanadium is a factor in controlling the $Na^+$, $K^+$-ATPase activity in this illness.

Vanadate(V) has been shown to block the activation of a progesterone receptor complex from avian oviduct[104]. Whether this effect involves phosphate metabolism is not known. A possible link between hormonal effects and the metabolism of vanadium in the reproductive physiology of rats has been investigated[165–167].

Some workers have reported interesting observations of vanadate effects on chromosome movement and spindle elongation in lysed mammalian cells[99] and on the assembly and function of 10 nm filaments and microtubles in syncytia cells, supporting the idea that they work together positioning nuclei and other organelles within the cell[168].

Topical application of vanadate(V) to rabbit and monkey eyes lowers the intraocular pressure because of a reduction in secretion of aqueous humor[169, 170]. This reduction has been attributed to inhibition of ciliary epithelium membrane $Na^+$, $K^+$-ATPase.

## *VI. Metabolism of Vanadium*

There is limited reliable information available on the distribution of vanadium in mammalian tissues. Because of the difficulty in accurately measuring vanadium in the ppb-ppm range in biological samples, much of the older literature is of questionable validity. Table 4 summarizes some reported values for vanadium concentration in several tissues. The molar concentrations of some tissues are of the same order of magnitude as many of the vanadium inhibitor constants given in Table 2 which lends support to the suggestion that vanadium could serve as an endogenous regulator of certain enzymes, provided, of course, that it is in the appropriate chemical form to do so.

In most studies of the metabolism of vanadium, the metal has been introduced into test animals by intravenous or intraperitoneal injection of various salts containing a $^{48}V$ tracer. Experiments in several laboratories, in which rats and mice were administered

**Table 4.** Concentration of vanadium in various tissues[a]

| Tissue | Concentration, μg/g, (μM)[b] | Reference |
|---|---|---|
| blood[c] | 0.011(0.19), 0.00077(0.014), 0.2(3.6), 0.02(0.36), 0.046(0.83) | 163, 201–204 |
| brain | 0.03(0.54) | 203 |
| dental enamel | <0.1(<1.8) | 203 |
| heart | 0.02(0.36)[d] | 204 |
| kidney | 0.03(0.54)[d] | 204 |
| liver | 0.05(0.9), 0.04(0.72)[d] | 203, 204 |
| lung | 0.10(1.8), 0.03(0.54)[d] | 203, 204 |
| lymph nodes | 0.40(7.2) | 203 |
| muscle | 0.01(0.18) | 203 |
| spleen | 0.02(0.36)[d] | 204 |
| testis | 0.2(3.6) | 203 |

[a] Human tissues except where noted

[b] First value listed is in units of μg vanadium per g wet weight of sample. Values for fluids have units μg/ml. The values in parentheses are the approximate concentrations expressed in micromolar units calculated by assuming all tissues have a density of approximately 1 g/ml

[c] Most studies indicate that little vanadium is present in red cells

[d] Dog

doses in the range 0.04 to 8 mg V per Kg body weight, lead to the conclusion that the preferential deposition of the metal in different tissues follows the approximate sequence: bone > kidney > liver > spleen > intestines, stomach, blood, muscle, testes, lung, brain[171–176].

Vanadium is deposited more slowly in bone than in other tissues and clears more slowly as well[172]. In this tissue it is possible that the metal is immobilized through a substitution of $VO_4^{3-}$ for $PO_4^{3-}$ in the apatite structure. Alternatively, $VO^{2+}$ has been shown to be adsorbed onto synthetic apatites, but this is relatively easily displaced[177,178].

The high levels (2 to 20 μM) immediately achieved in the kidney accounts for the diuretic and natriuretic effect of vanadate(V) mentioned earlier[19]. The reversal of this effect with time could come about by reduction of vanadate(V) to vanadyl(IV) and/or excretion of vanadium in the urine. About 25% of vanadium injected as vanadate(V) appears in the urine within 24 h[176]. The kidney reaches a maximum vanadium loading about 1.5 h after injection. The vanadium content of this organ then slowly drops over a period of days, long after diuresis and natriuresis has stopped. This observation suggests that vanadium is no longer present as vanadate(V). The vanadium excreted in the urine is associated with four fractions of different molecular weight, one of which may be transferrin[176]. We find that the addition of a pH 7.0, 0.05 M vanadate(V) solution to urine immediately results in the appearance of an EPR spectrum of a low molecular weight $VO^{2+}$ complex(es). Reduction therefore readily occurs in this fluid. It is therefore likely that the vanadium excreted via the kidney is present as $VO^{2+}$.

In the blood of rats, nearly 100% of the vanadium is associated with the plasma 0.5 h to 21 da after injection of vanadate(V)[176]. Uptake by red cells, which is a rapid process *in vitro*[174], occurs very little in the animal. Accumulation of vanadate(V) within the proximal tubules of the kidney[19] may be a more efficient process.

It is particularly significant that the distribution of vanadium among tissues or subcellular fractions is independent of whether the oxidation state of the vanadium salt administered is +3, +4, or +5[175, 179]. This result implies that all metabolic vanadium is converted to a common oxidation state in the circulation, probably to the $VO^{2+}$ ion, before being assimilated by most tissues, the kidney perhaps being an exception.

There is further evidence that V(V) may be converted to $VO^{2+}$ in serum, which then binds to the iron transport protein transferrin. This hypothesis is supported by the observation of Hopkins that vanadium, a few hours after injection, becomes associated with transferrin[8]. Sabbioni and Marafante[176] have also observed a small amount of vanadium bound to serum transferrin. However, in their experiments, and those of Hopkins, no precautions appear to have been taken to protect the vanadium from oxidation during the protein separation steps. Under these conditions, most $VO^{2+}$ associated with transferrin may be lost from the protein. The possibility that vanadate(V) binds to transferrin has not been investigated. If, in fact, most of the vanadium is bound to transferrin after a few hours, this could explain the failure of red cells to acquire much of the metal. These cells have relatively few receptor sites for the protein.

Liver and spleen, both with relatively high concentrations of vanadium (1 to 10 μM) after injection, are rich in the iron storage protein ferritin. These cells also have many receptor sites for transferrin. Whether the route of vanadium uptake by these cells is the same as that for iron remains to be determined. The vanadium in the cytosol of rat liver cells, which accounts for about 20% of the total vanadium uptake, has been shown to be associated with the ferritin and transferrin fractions[175]. It is well known that $VO^{2+}$ binds to transferrin[57, 180–183] and to ferritin[184] *in vitro*. Because these cells also have high levels of glutathione, it is likely that most or all of the vanadium is in the $VO^{2+}$ state. EPR observations confirm this[36, 58].

There is circumstantial evidence for a link between vanadium and iron metabolism (see Ref. 8). For example, in addition to the data discussed above, rats and chicks on low vanadium diets show elevated blood and bone iron levels and increased hematocrits, respectively. Sabbioni and coworkers[186] have shown that some vanadium introduced as $VO^{2+}$ into bovine milk becomes associated with lactoferrin and that vanadium bound to transferrin can exchange with ferritin when cellular components are present[187]. However, precautions against air oxidation of $VO^{2+}$ were also not taken in these experiments. Very low levels of vanadium were found associated with these proteins relative to their binding capacity.

$VO^{2+}$ binds stoichiometrically to the two specific $Fe^{3+}$ binding sites of human serum transferrin[57], rabbit serum transferrin[181], ovotransferrin[188, 189] and lactoferrin[190] *in vitro*. In addition, the unusual requirement that bicarbonate must bind to serum transferrin in order for $Fe^{3+}$ to bind has also been demonstrated for $VO^{2+}$[180, 183]. In contrast to these findings Sabbioni and coworkers claim[187] that eight g-atoms of V can associate with transferrin (no data given). If this is true, then, under the conditions of their experiments, most of the metal must be non-specifically bound to the protein.

The studies presented in this section suggest that vanadium-48 introduced into the circulation of rats finds its way into serum transferrin and ultimately into cells as $VO^{2+}$. The vanadium in the cytosol of rat liver cells is mostly found in transferrin and ferritin. Whether or not this is the normal physiological pathway for this metal has not been established by these experiments. The doses are administered over a short period of time and, in some cases, exceed by one or two orders of magnitude the concentration of V that would normally be present in serum. Moreover, precaution against air oxidation of vanadium, which may be critical in some of the experiments assigning vanadium to different fractions, was not taken.

It would be premature to conclude that vanadium and iron must share the same transport and storage proteins under normal physiological conditions. Both proteins are known to bind a variety of metal ions[191, 192]. In addition, since there is relatively little vanadium in the body and it is not needed in large amounts, a storage protein per se would not seem to be a necessary component of its biochemistry. To this extent at least, the metabolism of iron and vanadium seem to differ markedly. On the other hand it is a reasonable hypothesis that transferrin is the carrier protein for $VO^{2+}$ in blood. $VO^{2+}$ binds to this protein more strongly than to serum albumin[55], and there is evidence that transferrin may transport other metals besides iron[191]. A more definitive explanation of the metabolism of vanadium will have to await additional experimentation.

## E. Vanadyl(IV) EPR Spin Probe Studies

The specific binding of vanadium to a mammalian protein was first demonstrated in 1973 in studies with insulin[46]. This early work used a combination of X-ray crystallography, and infrared, visible, and EPR spectroscopy and competitive binding studies to show that $VO^{2+}$ bound at the same sites in the insulin crystal as normally occupied by the native metal $Zn^{2+}$. It was demonstrated that $VO^{2+}$ could likewise facilitate the formation of a protein hexamer in the solid state with the six subunits linked in part via two metal binding sites. Since that time, $VO^{2+}$ binding studies have been carried out on a relatively large number of proteins in which EPR spectroscopy was the major investigative tool employed. Other proteins investigated include carboxypeptidase A[205], carbonic anhydrase B[47, 193], serum albumin[53], thermolysin[194], hemoglobin[36, 58], staphylococcal nuclease[65], serum transferrin[180–183, 195, 196], conalbumin[188, 189], lactoferrin[190], and ferritin[184]. Some of the most fruitful studies have been with the non-heme iron proteins of storage and transport.

The thrust of most of these investigations was to gain information about metal sites with functional roles in proteins and to better understand the interactions between $VO^{2+}$ and proteins in general. $VO^{2+}$ EPR spin probe studies with proteins have been used to measure metal binding stoichiometry and association constants, determine the non-equivalence of different metal sites, establish which sites are functionally important, to identify some of the ligands, detect differences between solid state and solution samples, monitor changes in protein conformation, measure rotational correlation times, and determine the kinetics of metal ion-protein dissociation and exchange processes. In addition, spin probe measurements have been employed in studies of the polar surface of membranes, mineralization processes, RNA and DNA, humic and fulvic acids, liquid crystals and micelles, petroleum, and adsorption phenomena. These studies as well as the chemistry and spectroscopy of $VO^{2+}$ probes have been reviewed in detail elsewhere[21].

The ligands observed in vanadyl(IV) protein complexes include imidazole groups of histidine, carboxylate groups of glutamic and aspartic acid, phenolate groups of tyrosine, water, carbonate and other anions, ribonucleosides, and ribonucleotides. A protein complex in which sulfur of cysteine is a ligand has not been reported. There should be no obstacle to such a complex, since sulfur donor ligands are commonplace in the inorganic chemistry of $VO^{2+}$. $VO^{2+}$ has been shown to complex with the amino acid cysteine itself and with the methyl ester[38, 197], with sulfur in a $VO^{2+}$-transferrin-thioglycolate ternary complex[183], and with glutathione[206].

## F. Conclusion

Historically, vanadium has received little attention from biochemists. In the past four years, however, the interaction of this element with biological systems has become the focus of intense study in numerous laboratories throughout the world. Many biochemical and physiological effects of this essential metal are now known and new ones are being discovered almost monthly. Despite these recently discovered phenomena, the role of vanadium *in vivo* remains an open question. The potency of this element at extremely low concentrations, typical of those found in many tissues, is particularly relevant to a possible function. As is the case with most essential metals, vanadium in all probability will be shown to have more than one biological role.

Many of the biochemical observations reported to date can be understood in terms of the redox and coordination chemistry of vanadium. The major component of this element in soft tissues almost certainly exists as the $VO^{2+}$ ion, either complexed to proteins or to other biochelators ubiquitous in cells and fluids. Thus, one has an *in vivo* EPR probe of the coordination environment of the metal. Although the number of EPR investigations of cells has been limited to date, this is one promising avenue for future work. Sensitivity and background problems can largely be alleviated through computerized data acquisition and manipulation. There is now a reasonably large amount of EPR data on $VO^{2+}$-biomolecule complexes with which to compare the results obtained with cells. The conclusions derived from studies of vanadium metabolism would, in many instances, be put on a more firm basis if EPR measurements were also made. In favorable cases ENDOR spectroscopy might also be employed.

Except for studies of enzyme inhibition by vanadate(V), no other work appears to have been done on vanadium(V) binding to proteins. It would be useful to know whether orthovanadate(V) is specific for phosphate binding sites or is also capable of binding at other anion binding sites where it might also influence the properties of the protein. Moreover, the assumed trigonal bipyramidal complex of vanadate(V) with enzymes has never been actually demonstrated. While there is abundant data to indicate that $VO^{2+}$ binds to metal sites of proteins, the possibility that the vanadium(V) cation *cis*-$VO_2^+$ can also occupy these sites has not been explored. Such binding could be the first step in an in vivo pathway leading to the reduction of the vanadium(V) to form $VO^{2+}$-protein complexes.

Vanadium has earned its place among the trace metals worthy of biochemical investigation. In addition to being important in their own right, vanadium ions have provided the researcher with powerful tools for kinetically and spectroscopically probing complex cellular processes. We can look forward to new discoveries and insights into the biochemistry of this long neglected element as probe studies and the search for a function continue.

*Acknowledgment.* The author wishes to thank Mr. Donald A. Folajtar and Mr. Carl P. Thompson for critically reading the manuscript and Professor Kenneth Kustin for helpful suggestions on the section dealing with ascidians and Ms. Catherine Fisher for typing the manuscript. This work was supported by grant GM20194 through the National Institutes of Health.

## G. References


1. Weeks, M. E., Leicester, H. M.: Discovery of the Elements, pp 351–375, Easton, PA, Journal of Chemical Education, 1968
2. Oehme, F. W. (ed.): Toxicity of Heavy Metals in the Environment, Part 1, New York, Marcel Dekker, Inc., 1978
3. Oehme, F. W. (ed.): Toxicity of Heavy Metals in the Environment, Part 2, New York, Marcel Dekker, Inc., 1979
4. Hopkins, L. L.: Geochem. Environ., Vol II, 93 (1977)
5. Goldberg, E. D.: Minor Elements in Seawater in: Chemical Oceanography, Vol. 1, (eds.) Riley, J. P., Skirrow, G., p 163, New York, Academic Press, 1965
6. Yen, T. F.: The Role of Trace Metals in Petroleum, Ann Arbor Mich., Ann Arbor Publishers, 1975
7. Maylotte, D. H. et al.: Science *214,* 554 (1981)
8. Hopkins, L. L.: Essentiality and Function of Vanadium, in: Trace Element Metabolism in Animals, Proceedings International Symposium on Trace Element Metabolism in Animals, p 397, Baltimore, University Park Press, 1974
9. McBride, M. B.: Can. J. Soil Sci. *60,* 145 (1980)
10. Templeton III, G. D., Chasteen, N. D.: Geochim. Cosmochim. Acta *44,* 741 (1980) and references therein
11. Cheshire, M. V. et al.: Geochim. Cosmochim. Acta *41,* 1131 (1977)
12. Hopkins, Jr., L. L., Mohr, H. E.: The Bioessentiality of Vanadium in: Newer Trace Elements in Nutrition (eds.) Mertz, W. and Cornatzer, W. E., p 195, New York, Marcel Dekker, 1971
13. Schwarz, K., Milne, D. B.: Science *174,* 426 (1971)
14. Nielsen, F. H.: Adv. Nutr. Res. *3,* 157 (1980)
15. Vouk, V. B.: Vanadium in: Handbook on the Toxicology of Metals (eds.) Frieberg, L., Nordberg, G. F., Vouk, V. B., p 659, Amsterdam–New York–Oxford, Elsevier/North Holland Biomedical Press, 1979
16. Macara, I. G.: TIBS, April, 92 (1980)
17. Pope, M. T., Still, E. R., Williams, R. J. P.: A Comparison between the Chemistry and Biochemistry of Molybdenum and Related Elements in: Molybdenum and Molybdenum containing Enzymes, pp 1–40, New York, Pergamon Press, 1980
18. Simons, T. J. B.: Nature (London) *281,* 337 (1979)
19. Grantham, J. J.: Am. J. Physiol. *239,* 97 (1980)
20. Biggs, W. R., Swinehart, J. H.: Vanadium in Selected Biological Materials in: Metal Ions in Biological Systems, Vol. 6 (ed.) Sigel, H., pp 142–196, New York, Marcel Dekker, Inc., 1976
21. Chasteen, N. D.: Vanadyl(IV) EPR Spin Probes, Inorganic and Biochemical Aspects in: Biological Magnetic Resonance, Vol. 3 (eds.) Berliner, L., Reuben, J., p 53, New York, Plenum Press, 1981
22. Tullius, T. D. et al.: J. Am. Chem. Soc. *102,* 5670 (1980)
23. Pope, M. T., Dale, B. W.: Quart. Rev. (London) *22,* 527 (1968)
24. Borgen, O., Mahmoud, M. R., Skauvik, I.: Acta Chem. Scand. *A31,* 329 (1967)
25. Amos, L. W., Sawyer, D. T.: Inorg. Chem. *11,* 2692 (1972)
26. Lindquist, R. N., Lynn, J. L., Lienhard, G. E.: J. Am. Chem. Soc. *95,* 8762 (1973)
27. Choate, G., Mansour, T. E.: J. Biol. Chem. *254,* 11457 (1979)
28. DeMaster, E. G., Mitchell, R. A.: Biochemistry *12,* 3616 (1973)
29. Hayayeb, M. A., Hileman Jr., O. E.: Can. J. Chem. *58,* 2255 (1980)
30. Baes, C. F., Mesmer, R. E.: The Hydrolysis of Cations, pp 197–210, New York, Wiley-Interscience, 1976
31. Francavilla, J., Chasteen, N. D.: Inorg. Chem. *14,* 2860 (1975)
32. Iannuzzi, M. M., Rieger, P. H.: Inorg. Chem. *14,* 2895 (1975)
33. Schechter, Y., Karlish, S. J. D.: Nature (London), *284,* 556 (1980)
34. Kragten, J.: Atlas of Metal-Ligand Equilibria in Aqueous Solution, Chap 42, Chichester, Ellis Horwood Limited, 1978
35. Kustin, K., Toppen, D. L.: Inorg. Chem. *12,* 1404 (1973)
36. Macara, I. G., Kustin, K., Cantley Jr., L. C.: Biochim. Biophys. Acta *629,* 95 (1980)

37. Cantley Jr., L. C., Ferguson, J. H., Kustin, K.: J. Am. Chem. Soc. *100,* 5210 (1978)
38. Sakurai, H., Shimomura, S., Ishizu, K.: Inorg. Chim. Acta *55,* 67L (1981)
39. Erdman, E., et al.: Basic Res. Cardiol. *75,* 460 (1980)
40. Vyskocil, F., Teisinger, J.: Nature (London) *286,* 516 (1980)
41. Ramasarma, T., MacKellar, W. C., Crane, F. L.: Biochim. Biophys. Acta *646,* 88 (1981)
42. Erdman, E., et al.: Nature (London) *282,* 335 (1979)
43. Woltermann, G. M., Belford, R. L., Haight, G. P.: Inorg. Chem. *16,* 2985 (1977)
44. Imamura, T., et al.: J. Inorg. Biochem. *11,* 241 (1979)
45. Woltermann, G. M., Scott, R. A., Haight, G. P.: J. Am. Chem. Soc. *96,* 7569 (1974)
46. Chasteen, N. D., et al.: J. Am. Chem. Soc. *95,* 1301 (1973)
47. Bertini, I. et al.: Inorg. Chim. Acta *36,* 9 (1979)
48. O'Donnell, S. E., Pope, M. T.: J. Chem. Soc. Dalton 2290 (1976)
49. White, L. K., Chasteen, N. D.: J. Phys. Chem. *83,* 279 (1979)
50. Johnson, J. L., Cohen, H. J., Rajagopalan, K. V.: Biochem. Biophys. Res. Commun. *56,* 940 (1974)
51. Sakurai, H. et al.: Biochem. Biophys. Res. Commun. *96,* 293 (1980)
52. Dickson, F. E. et al.: Anal. Chem. *44,* 978 (1972)
53. Boucher, L. J., Tynan, E. C., Yen, T. F.: Spectral Properties of Oxovanadium(IV) Complexes. IV Correlation of ESR Spectra with Ligand Type in: Electron Spin Resonance of Metal Complexes, (ed.) Yen, T. F., pp 111–130, New York, Plenum Press, 1969
54. Campbell, R. F., Freed, J. H.: J. Phys. Chem. *84,* 2668 (1980)
55. Chasteen, N. D., Francavilla, J.: J. Phys. Chem. *80,* 897 (1976)
56. Fitzgerald, J., Chasteen, N. D.: Anal. Biochem. *60,* 170 (1974)
57. Cannon, J. C., Chasteen, N. D.: Biochemistry *14,* 4573 (1975)
58. Cantley, Jr., L. C., Aisen, P.: J. Biol. Chem. *254,* 1781 (1979)
59. Mulks, C. F., van Willigen, H.: J. Phys. Chem. *85,* 1220 (1981)
60. van Willigen, H.: J. Magn. Reson. *39,* 37 (1980)
61. van Willigen, H.: Chem. Phys. Lett. *65,* 490 (1979)
62. van Willigen, H., Mulks, C. F., Bouhaouss, A., Ferhat, M., Roufosse, A. H.: J. Am. Chem. Soc. 102, 4846 (1980)
63. Dikanov, V. F., Yudanov, V. F., Tsvetkov, Y. D.: J. Magn. Reson. *34,* 631 (1979)
64. Mims, W. B., Peisach, J.: Electron Spin Echo Spectroscopy and the Study of Metalloproteins in: Biological Magnetic Resonance, Vol. 3, (eds.) Berliner, L. J., Reuben, J., pp 213–263, New York, Plenum Press, 1981
65. Tucker, P. W.: Chemical and Crystallographic Studies of Staphylococcal Nuclease: Ph. D. Thesis Texas A & M University, College Station, Texas, 1975
66. Tucker, P. W., Hazen, Jr., E. E., Cotton, F. A.: Mol. Cell. Biochem. *23,* 67 (1979)
67. Berger, S. L., Birkenmeier, C. S.: Biochemistry *18,* 5143 (1979)
68. Lopez, V., Stevens, T., Lindquist, R. N.: Arch. Biochem. Biophys. *175,* 31 (1976)
69. Seargeant, L. E., Stinson, R. A.: Biochem. J. *181,* 247 (1979)
70. VanEtten, R. L., Waymack, P. P., Rehkop, D. M.: J. Am. Chem. Soc. *96,* 6782 (1974)
71. Cantley, Jr., L. C. et al.: J. Biol. Chem. *252,* 7421 (1977)
72. Stryer, L.: Biochemistry, Chap 36, San Francisco, W. H. Freeman and Co., 1981[2]
73. Sachs, G.: J. Lab. Clin. Med. *96,* 379 (1980)
74. Cantley, Jr., L. C., Resh, M. D., Guidotti, G.: Nature (London) *272,* 552 (1978)
75. Beaugé, L. A., Glynn, I. M.: Nature (London) *272,* 551 (1978)
76. Byskocil, F., Teisinger, J., Dlouha, H.: Biochem. Biophys. Res. Commun. *100,* 982 (1981)
77. Wallick, E. T., Lane, L. K., Schwartz, A.: J. Biol. Chem. *254,* 8107 (1979)
78. Smith, R. L., Zinn, K., Cantley, Jr., L. C.: J. Biol. Chem. *255,* 9852 (1980)
79. Karlish, S. J. D., Beaugé, L. A., Glynn, I. M.: Nature (London) *282,* 333 (1979)
80. Hobbs, A. S., Fruelich, J. P., Albers, R. W.: J. Biol. Chem. *255,* 5724 (1980)
81. Cantley, Jr., L. C., Cantley, L. G., and Josephson, L.: J. Biol. Chem. *253,* 7361 (1978)
82. Bond, G. H., Hudgins, P. M.: Biochemistry *18,* 325 (1979)
83. Kaerlish, S. J. D., Yates, D. W., Glynn, I. M.: Biochim. Biophys. Acta *525,* 230 (1978)
84. Hansen, O.: Basic Res. Cardiol. *75,* 455 (1980)
85. Agency of Industrial Sciences and Technology. Patent Application 79/36, 658 (1979) CA94: 79893

86. Rossi, J. P., Garrahan, P. J., Rega, A. F.: Biochim. Biophys. Acta *648,* 145 (1981)
87. Bond, G., Hudgins, P. M.: Biochim. Biophys. Acta *600,* 781 (1980)
88. Barrabin, H., Garrahan, J., Rega, A. F.: Biochim. Biophys. Acta *600,* 796 (1980)
89. Josephson, L., Cantley, Jr., L. C.: Biochemistry *16,* 4572 (1977)
90. O'Neal, S. G. et al.: Biochem. Biophys. Res. Commun. *89,* 845 (1979)
91. Hagenmeyer, A. et al.: Basic Res. Cardiol. *75,* 452 (1980)
92. Bowman, B. J., Slayman, C. W.: J. Biol. Chem. *254,* 2928 (1979)
93. Borst-Pauwels, G. W. F. H., Peters, P. H. J.: Biochim. Biophsy. Acta *642,* 173 (1981)
94. Willsky, G. R.: J. Biol. Chem. *254,* 3326 (1979)
95. Goodno, C. C.: Proc. Natl. Acad. Sci. USA *76,* 2620 (1979)
96. Gibbons, I. R. et al.: Proc. Natl. Acad. Sci. USA *75,* 2220 (1978)
97. Shimizu, T.: Biochemistry *20,* 4347 (1981)
98. Kobayashi, T. et al.: Biochem. Biophys. Res. Commun. *81,* 1313 (1978)
99. Cande, W. Z., Wolniak, S. M.: J. Cell. Biol. *79,* 573 (1978)
100. Nelson, M. T., Blaustein, M. P.: Nature (London) *289,* 314 (1981)
101. Dipolo, R., Beaugé, L.: Biochim. Biophys. Acta *645,* 229 (1981)
102. Dubyak, G. R., Kleinzeller, A.: J. Biol. Chem. *255,* 5306 (1980)
103. Hori, C., Oka, T.: Biochim. Biophys. Acta *610,* 235 (1980)
104. Carpenter, G.: Biochem. Biophys. Res. Commun. *102,* 1115 (1981)
105. Seglen, P. O., Gordon, P. B.: J. Biol. Chem. *256,* 7699 (1981)
106. Bertrand, D.: Bull. Soc. Chim. Biol. *25,* 194 (1943)
107. Kneifel, H., Bayer, E.: Angew. Chem. Int. Ed. Engl. *12,* 508 (1973)
108. Lancashire, Robert J.: Educ, Chem. *17,* 38 (1980)
109. Kustin, K., Ladd, K. V., McLeod, G. C.: J. Gen. Physiol. *101,* 65 (1975)
110. Carlson, R. M. K.: Proc. Natl. Acad. Sci. USA *72,* 2217 (1975)
111. Swinehart, J. H. et al.: Biol. Bull. *146,* 302 (1974)
112. Kustin, K. et al.: Biol. Bull. *150,* 426 (1976)
113. Dingley, A. J., Kustin, K., Macara, I. G., McLeod, G. C.: Biochim. Biophys. Acta. *649,* 493 (1981)
114. Macara, I. G., McLeod, G. C., Kustin, K.: Biochem. J. *181,* 457 (1979)
115. Macara, I. G., McLeod, G. C., Kustin, K.: Comp. Biochem. Biophys. *63B,* 299 (1979)
116. Gilbert, K., Kustin, K., McLeod, G. C.: J. Cell Physiol.: *93,* 309 (1977)
117. Macara, I. G., McLeod, G. C., Kustin, K.: Comp. Biochem. Physiol. *62A,* 821 (1979)
118. Stoecker, D.: Ecology *61*, 1327 (1980)
119. Climent, F. et al.: Biochim. Biophys. Res. Commun. *101,* 570 (1981)
120. Pai, E. F. et al.: J. Mol. Biol. *114,* 37 (1977)
121. Carreras, J. et al.: Biochem. Biophys. Res. Commun. *96,* 1267 (1980)
122. Layne, P. P., Najjar, V. A.: Proc. Natl. Acad. Sci. USA *76,* 5010 (1979)
123. Catalán, R. E., Martínez, A. M., Aragones, M. D.: Biochem. Biophys. Res. Commun. *96,* 672 (1980)
124. Day, H. et al.: J. Lab. clin. Med. *96*, 382 (1980)
125. Balfour, W. E., Grantham, J. J., Glynn, I. M.: Nature *275,* 768 (1978)
126. Ehrenspeck, G.: Biochim. Biophys. Acta *601,* 427 (1980)
127. Fanestil, D. D.: Experimentia *36,* 1045 (1980)
128. Beauwens, R.: J. Physiol. (London) *310,* 293 (1981)
129. Kelly, K. F., Bell, M. V., Sargent, J. R.: Biochim. Biophys. Acta *644,* 376 (1981)
130. Schwabe, U. et al.: Nature (London) *277,* 143 (1979)
131. Borchard, U. et al.: J. Cardiovasc. Pharmacol. *3,* 510 (1981)
132. Peterson, J. W.: Biochem. Biophys. Res. Commun. *95,* 1846 (1980)
133. Schwartz, A. et al.: Basic Res. Cardiol. *75,* 444 (1980) and references therein
134. Larsen, J. A., Thomsen, O. O.: Basic Res. Cardiol. *75,* 428 (1980)
135. Steffen, R. P. et al.: Hypertension (Dallas), *3,* 173 (1981) CA95: 1177m
136. Takeda, K. et al.: Naunyn-Schmiedebergs Arch. Pharmacol. *314,* 161 (1980)
137. Borchard, U. et al.: Basic Res. Cardio. *75,* 423 (1980)
138. Inciarte, D. J. et al.: Am. J. Physiol. *239,* H47 (1980)
139. Solaro, R. J. et al.: J. Cardiovasc. Pharmacol. *2,* 445 (1980)
140. Scholz, H. et al.: Basic Res. Cardiol. *75,* 418 (1980)

141. Werden, K. et al.: Basic Res. Cardiol. *75,* 466 (1980)
142. Schmitz, W. et al.: Basic Res. Cardiol. *75,* 438 (1980)
143. Rapp, J. P.: Hypertension (Dallas) *3,* 168 (1981) CA95: 1176k
144. Brueckner, R. et al.: Naungn-Schmiedeberg's Arch. Pharmacol. *315,* 147 (1980)
145. Larsen, J. A., Thomsen, O.: Acta Physiol. Scand. *110,* 367 (1980)
146. Meisch, H. U., Bielig, H. J.: Basic Res. Cardiol. *75,* 413 (1980)
147. Sargent, J. R. et al.: Comp. Biochem. Physiol. C, 66C (1980)
148. Borchard, U. et al.: Nature (London) *279,* 339 (1979)
149. Myers, T. D., Boerth, R. C.: Biochem. Biophys. Res. Commun. *96,* 39 (1980)
150. Herzig, J. W. et al.: Biochim. Biophys. Acta *672,* 191 (1981)
151. Tritthart, H. A. et al.: Basic Res. Cardiol. *75,* 472 (1980)
152. Garcia, A. et al.: Eur. J. Pharmacol. *70,* 17 (1981)
153. Fein, A., Corson, D. W.: Science *212,* 555 (1981)
154. Burns, R. C., Fuchsman, W. H., Hardy, R. W. F.: Biochem. Biophys. Res. Commun. *42,* 353 (1971)
155. Beneman, J. R. et al.: Biochim. Biophys. Acta *264,* 25 (1972)
156. Lee, K. Y. et al.: J. Biol. Chem. *249,* 3953 (1974)
157. Ramadoss, C. S.: Z. Naturforsch., C: Biosci. *35C,* 702 (1980)
158. Buczek, J.: Acta Soc. Bot. Pol. *49,* 269 (1980)
159. Menon, A. S., et al.: FEBS Lett. *114,* 139 (1980)
160. Naylor, G. J., Smith, A. H. W.: IRCS Med. Sci.: Libr. Compen. *8,* 446 (1980)
161. Earnshaw, K. R. et al.: Pharmacol. Res. Commun. *13,* 323 (1981)
162. Naylor, G. J.: Neuropharmacology *19,* 1233 (1980)
163. Dick, D. A. T. et al.: J. Physiol. (London) *310,* 24P (1981)
164. Nishigori, H. et al.: Arch. Biochem. Biophys. *203,* 600 (1980)
165. Peabody, R. A. et al.: Proc. Soc. Exp. Biol. Med. *165,* 349 (1980)
166. Peabody, R. A. et al.: Trace Subst. Environ. Health XI, 297 (1977)
167. Peabody, R. A. et al.: Trace Subst. Environ. Health X, 441 (1976)
168. Wang, E., Choppin, P. W.: Proc. Natl. Acad. Sci. USA *78,* 2367 (1981)
169. Becker, B.: Invest. Ophthalmol. Visual Sci. *19,* 1156 (1980)
170. Kupin, T. et al.: Invest. Ophthalmol. Visual Sci. *19,* 1360 (1980)
171. Sharma, R. P. et al.: J. Toxicol. Environ. Health *6,* 45 (1980)
172. Parker, R. D. R. et al.: Arch Environ. Contam. Toxicol. *9,* 393 (1980)
173. Ordzhonikidze, E. K. et al.: J. Hyg., Epidemiol., Microbiol., Immunaol. *24,* 377 (1980)
174. Oberg, S. G. et al.: Trace Subst. Environ. Health *13,* 432 (1979)
175. Sabbioni, E. et al.: Bioinorg. Chem. *8,* 503 (1978)
176. Sabbioni, E., Marafante, E.: Bioinorg. Chem. *9,* 389 (1978)
177. Peckauskas, R. A., Pullman, I.: Biopolymers *15,* 569 (1976)
178. Peckauskas, R. A., Pullman, I., Termine, J. D.: Biopolymers *16,* 199 (1977)
179. Hopkins, L. L., Tilton, B. E.: Am. J. Physiol. *211,* 169 (1966)
180. Chasteen, N. D., White, L. K., Campbell, R. F.: Biochemistry *16,* 363 (1977)
181. Chasteen, N. D. et al.: Vanadyl(IV) Labelled Transferrin: An Overview in: Proteins of Iron Metabolism, (eds.) Brown, E. B. et al., pp 187–195, New York, Grune and Stratton, 1977
182. Cannon, J. C., Chasteen, N. D.: The Distinction between Metal Binding Sites in Vanadyl Transferrin Complexes in: Proteins of Iron Storage and Transport in Biochemistry and Medicine, (ed.) Crichton, R. R., Amsterdam, North Holland Press, 1975
183. Campbell, R. F., Chasteen, N. D.: J. Biol. Chem. *252,* 5996 (1977)
184. Chasteen, N. D., Theil, E. C.: J. Biol. Chem. *257,* in press
185. Snipes, W., Gordy, W.: J. Chem. Phys. *41,* 3661 (1964)
186. Sabbioni, E. and Rade, J.: Toxicol. Lett. *5,* 381 (1980)
187. Sabbioni, E., Rade, J., Bertolero, F.: J. Inorg. Biochem. *12,* 293 (1980)
188. Casey, J. D., Chasteen, N. D.: J. Inorg. Biochem. *13,* 111 (1980)
189. Casey, J. D., Chasteen, N. D.: J. Inorg. Biochem. *13,* 127 (1980)
190. Carmichael, A., Vincent, J. S.: FEBS Lett. *105,* 349 (1979)
191. Chasteen, N. D.: Coord. Chem. Rev. *22,* 1 (1977)
192. Clegg, G. A. et al.: Prog. Biophys. Molec. Biol. *36,* 53 (1981)
193. Fitzgerald, J., Chasteen, N. D.: Biochemistry *13,* 4338 (1974)

194. Harrington, J. K.: ESR Study of Vanadyl(IV) Labeled Thermolysin, M. S. Thesis, University of Texas at El Paso, El Paso, Texas, 1978
195. Harris, D. C.: Biochemistry *16,* 560 (1977)
196. Harris, D. C., Gelb, M. H.: Biochim. Biophys. Acta *623,* 1 (1980)
197. Sakurai, H. et al.: Inorg. Chim. Acta *46,* L119 (1980)
198. Kasuga, M., Karlsson, F. A., Kahn, C. R.: Science *215,* 185 (1982)
199. Goodno, C. C., Taylor, E. E.: Proc. Natl. Acad. Sci. USA *79,* 21 (1982)
200. Quist, E. E., Hokin, L. E.: Biochim. Biophys. Acta *511,* 202 (1978)
201. Allen, R. O., Steinnes, E.: Anal. Chem. *50,* 1553 (1978)
202. Agrawal, Y. K., Sant, M. S.: Bioinorg. Chem. *9,* 369 (1978)
203. Underwood, E. J.: Trace Elements in Human and Animal Nutrition, p 389, New York, Academic Press, 1977[4] and references therein
204. Hayoshi, M.: Bull. Environm. Contam. Toxicol. *25,* 246 (1980)
205. DeKoch, R. J. et al.: Biochemistry *13,* 4347 (1974)

# Vanadium and Other Metal Ions in the Physiological Ecology of Marine Organisms

**Kenneth Kustin[1] and Guy C. McLeod[2] with the assistance of Thomas R. Gilbert[3] and Le Baron R. Briggs, 4th[4]**

1 Department of Chemistry, Brandeis University, Waltham, Massachusetts 02254, U.S.A.
2 New England Aquarium, Central Wharf, Boston, Massachusetts 02110, U.S.A.
3 Department of Chemistry, Northeastern University, Boston, Massachusetts 02115, U.S.A.
4 Department of Biology, Northeastern University, Boston, Massachusetts 02115, U.S.A.

Both natural and polluting metal ions are taken up by marine organisms. For many metal, and metal-containing ions, mechanisms exist which allow these organisms to discriminate between essential and non-essential ions, and reject the latter. In this article we examine the accumulation mechanisms and effects of metal and metal-containing ions on bivalve molluscs and tunicates (sea squirts). Special emphasis is put on the internal or physiological environment of metal ions in marine organisms. A molecular mechanism based on experimental studies is presented to explain the extraordinary accumulation of vanadium in tunicate blood cells called vanadocytes. Vanadium in the form of vanadate anion is transported into the blood cells by passage through anionic channels. Inside the cell it is reduced to cationic V (III) and V (IV) ions which are trapped inside.

# I. Introduction

Marine organisms contend with both natural and polluting metal ions in two quite dissimilar environments. Uptake of metal ions from the external environment occurs with little, if any, selection or rejection. Therefore, once the organism/environment barrier is penetrated, a living organism may be faced with a juxtaposition of similar ions, the slight differences between which confer essentiality on some, and toxicity on others. An earlier review considered only the interactions of metal ions with marine organisms in the external, sea water environment[1)]. In the present treatment the concept of environment is extended to the internal, or physiological, interactions of metal ions in marine organisms.

Previously, reviews of this type have centered on the organism, rather than the contribution of the metal ions from the external environment[2)]. In this review, selected trace elements are placed in focus and followed from their sources through sea water, into blood and other tissues, finally into specific cells. The micro-environment of a trace metal is seen to be important; for example, the presence or absence of other ions will determine the conditions of competition in uptake by marine organisms. In the inner environment of living organisms, the presence or absence of chelating, reducing or oxidizing agents will determine whether assimilation or rapid rejection will be the fate of a particular ion.

After a treatment of general aspects of metals in the marine environment, specific target organisms, mainly bivalves (clams, mussels, oysters) and tunicates are considered. The latter class of marine organisms, commonly referred to as sea squirts, allows us to trace the pathway of a specific element, vanadium, from outer to inner environment.

# II. Assimilation of Metal Ions

Intracellular accumulation of metal ions may entail three separate processes:

1) Complex formation in the extracellular environment (for all ions other than $Na^+$, $K^+$, $Mg^{2+}$ and $Ca^{2+}$.)
2) Transport of the ion through a membrane barrier.
3) Binding or oxidation-reduction of the metal ion followed by binding inside the cell.

In the absence of an active ion-pumping mechanism, the last step is necessary to ensure accumulation, rather than merely transmembrane equilibration. These processes occur at the molecular level, and will be examined beginning with complex formation.

## 1. *Kinetics of Metal Ion Binding*

The general features of metal complex formation become apparent if metal ions are grouped according to ease of ligand substitution[3)]. A metal cation in water is solvated, and complex formation is the replacement (substitution) of a bound water molecule by one or more incoming ligands, or incoming ligand dentate sites. Thus, the rate constant for water exchange on the metal ion, $k_0$ ($s^{-1}$), largely determines the rate of this reaction. On this basis metal ions can be sorted into three categories.

(I), $k_0 > 10^8 s^{-1}$ (alkali metals, $Ca^{2+}$, $Sr^{2+}$, $Ba^{2+}$, $Cu^{2+}$, $Cr^{2+}$, $Zn^{2+}$, $Cd^{2+}$, $Hg^{2+}$).
(II), $10^2 < k_0 < 10^8 s^{-1}$ ($Mg^{2+}$, $Ga^{3+}$, $In^{3+}$, rare earth ions, 1st block transition metal divalent ions (except $Cu^{2+}$, $Cr^{2+}$), and oxoions such as $VO^{2+}$, $VO_2^+$).
(III), $k_0 < 10^2 s^{-1}$ ($Be^{2+}$, $Al^{3+}$, $Fe^{3+}$, $Cr^{3+}$, $Co^{3+}$).

Complex formation is at least a two-step process[3], with each step usually being well separated in time. The formation of an ion-pair between incoming ligand and cation neutralizes charge, but leaves the metal ion coordination shell intact. Upon release of an inner sphere water molecule, the ligand in the ion-pair can capture the vacant site. If the rate-determining step is cleavage of the metal-water bond, the mechanism is dissociative (D). In this case, the water exchange rate constant is characteristic of the metal ion, and ligand properties have essentially no effect on the rate of substitution. Category II ions exhibit this feature. If the rate-determining step is attachment of ligand to metal in an activated complex of expanded coordination number (compared with the aquo ion), the mechanism is associative (A). In this type of mechanism, the rate of complex formation is influenced by ligand properties. Ions in categories I and III follow this mechanism, although for different reasons.

Category I ions are so labile that the two steps of ion-pair formation and water expulsion merge. Ligand properties can then influence penetration into the metal ion's inner coordination shell. Category III ions are relatively inert, and hydrolysis (OH-substitution) becomes competitive with ligand substitution. This mechanism, by inserting negative charge into the inner coordination shell, assists expulsion of bound dipolar water molecules. The kinetics are then more complex than for categories I and II, leading to the many effects observed in the classical studies of Cr(III), Co(III), Pt(IV), etc., complex formation[4)]. Processes intermediate between A and D, termed interchange ($I_A$ or $I_D$), are also possible[4)], and may have been observed with some ions[5,6)]. These considerations pertain to the outer ecology of marine organisms, but a different situation arises in the internal environment.

Once a metal ion has penetrated the organism/environment interface, it soon appears in the blood. The homogeneous aqueous phase of blood, called plasma, is a medium of ionic strength about 0.17 M (mainly NaCl) and average pH about 7.4[7)]. Sea water has about twice the ionic strength, 0.35 M (mainly NaCl), and a higher average pH, about 8, than blood plasma[8)]. Comparisons between levels of dissolved organic material are less certain, due to the nonuniform distribution in sea water.

The concentration of dissolved organic matter in sea water is diffucult to specify because of variations due to location, season and depth. The best estimates place the total dissolved organic material in the 2–3 mmol of carbon range[8)]. Due to the dispersive, oxidizing conditions in the upper levels of sea water, the main organic species are monomeric amino acids, fatty acids and carbohydrates. All of these compounds are good chelators; however, the high dilution of sea water shifts the equilibrium towards free ions and ligands. In blood plasma, by contrast, dissolved organic substances are
(a) of more significant concentration, about 20 mmol, and
(b) are present to a large extent as macromolecules, namely proteins, in addition to being present as organic acids[9)]. Carbohydrates are present, but at lower levels.

Intracellular fluids contain higher concentrations of dissolved organic substances, mainly proteinaceous, at about 50 mmol, and although the ionic strength and pH do not differ appreciably from that of plasma, two important differences exist:
(a) $[K^+] \gg [Na^+]$, and (b) [phosphate species] $\gg$ [carbonate species].

The consequences for metal ion complex formation in making the transit from sea water to cell is the shift in equilibrium from simple aquo metal ions to complexes. Thus, transfer of metal ions is no longer a matter of substitution reactions, but rather of metal exchange between two complexes, or between complex and free ligand. In fact, elevated concentrations of free metal ions in serum or cell fluids produce deleterious abnormalities such as Wilson's disease in humans[10]. Thus, in addition to the metal complex formation reaction (1), we should consider ligand exchange (2) and double ligand exchange (3).

$$M + L \rightleftharpoons ML \tag{1}$$
$$ML + L' \rightleftharpoons ML' + L \tag{2}$$
$$ML + M'L' \rightleftharpoons ML' + M'L \tag{3}$$

In these reactions M and M′ represent metal ions or oxoions; L and L′ represent ligands which, for reactions (2) and (3), are usually multidentate. (Other reactions, such as single ligand exchange between two metal ions, will not be as important in the internal environment, and will not be considered here).

Chelation confers added stability to a metal complex compared with monodentate ligands of the same structural type; *viz.* ethylendiamine binding vs. binding of two ammonia molecules[11]. Kinetically, the stability results from extremely low-valued dissociation rate constants for breaking chelate rings. Therefore, when L in reaction (2) is a chelate, this process is usually relatively slow; however, under certain circumstances this type of reaction can be surprisingly rapid[12]. Thus, if the ring-opening can be accelerated, or if the incoming ligand can gain a position in the inner coordination shell, interchange can be rapid. In the case of reaction (3), small amounts of free ligand can catalyze the reaction, leading to coordination chain reactions.

The reactions discussed entail coordination of simple aquo ions such as $Cu^{2+}$, $Mg^{2+}$, etc. Metal-containing oxo-anions such as $VO_4^{3-}$, $MoO_4^{2-}$ and $WO_4^{2-}$ also form stable complexes, especially with cis-diols[13]. These reactions are quite rapid; vanadate, for example, has bimolecular complex formation rate constants on the order of $10^4\ M^{-1}\ s^{-1}$[14].

The oxoanions cited above are tetrahedral in their unprotonated forms[15], but protonation may cause expansion of the inner coordination shell, resulting in six-fold octahedral coordination[16]. Although some ambiguity exists on the exact structure of protonated oxoions in solution, no doubts exist as to the bonding in the complexes: these are octahedral[15].

Three features characterize the kinetics of oxoanion complex formation.
(1) The protonated oxoanion reacts faster than the unprotonated oxoanion.
(2) The more basic the ligand, the faster the reaction with the protonated oxoanion.
(3) For the unprotonated oxoanion, the least basic ligand is most reactive[17, 18].

While few ligand exchange reactions have been studied, those reported are slow[19]. Although the process of ligand exchange could be rate determining in cellular accumulation, in at least one system – vanadate uptake by human erythrocytes – the overall process is not limited by this reaction[20].

## *2. Accumulation Mechanisms*

Understanding how ions are transported through membranes is fundamental to biology[9]. No attempt will be made here to review the vast literature on this subject. Instead, we will concentrate on distinguishing the separate mechanisms by which cells tend to select and accumulate metal ions.

A few cations are transported vectorially into and out of cells, leading to transmembrane concentration gradients. This type of transport, called active transport, is important for $H^+$, $Na^+$, $K^+$, $Ca^{2+}$ and (possibly) $Mg^{2+}$. Specific enzymes are found in membranes which couple the hydrolysis of ATP to ion transport, and thereby achieve active pumping[21]. The activities of humans can seriously affect this type of transport.

Ions not normally present in the environment, that are similar in charge and size to the natural substrate ions, can interfere with active transport. Ions such as $Cd^{2+}$ and $Ba^{2+}$ are not normally abundant enough in sea water to pose a threat, but can interfere with $Ca^{2+}$ and $Mg^{2+}$ uptake in polluted waters.

Other ions, including many that are essential, are first chelated, and as neutral or negatively charged complexes, are transported into cells. In this way, $Cr^{3+}$ could penetrate the organism/environment barrier via the same route taken by $Fe^{3+}$. Oxoanions are negatively charged; as analogues of phosphate and sulfate these ions enter organisms through anionic channels in membranes[22]. A proven example of oxoanion transport through anionic channels is afforded by the accumulation of vanadium in tunicates, which is discussed below. Before doing so, we trace the pathway of metal ions from external to internal environment, and consider the problem of metal ion toxicity.

# III. Physical and Chemical Forms of Metal Ions in Sea Water

Laboratory studies of metal ion-ligand or metal ion-organism interactions are carried out under controlled conditions. In the marine evironment, however, organisms encounter metal ions under a variety of conditions. Most trace elements rarely occur in sea water as simple hydrated ions, in complexes with the major marine anions ($Cl^-$, $HCO_3^-$, $SO_4^{2-}$), or in complexes with dissolved organic compounds. Although these forms can serve as metal ion sources to organisms, as in the case of vanadium and tunicates, other sources are more abundant.

The most likely sources of metal ions for marine organisms are the colloidal or particulate moieties such as plankton, organic detritus, and clay minerals. The latter may be coated with iron or manganese oxides which are efficient scavengers of metals in slightly alkaline media.

Not unexpectedly, trace metal cycling in estuarine environments and the offshore migration of metals are dominated by the transport of suspended particulate forms[23]. The importance of these phases in the aquatic chemistries of trace metals has become clear only recently. One reason for this is that most of the dissolved metal fractions previously measured in environmental waters were analytical artifacts caused by sample contamination.

One mechanism for offshore migration involves the resuspension of fine-grained estuarine sediments under tidal action and high wind stress. A given particle probably undergoes repeated resuspension and deposition before finally settling into a bathymetric low on the continental shelf[24)], or in the abyssal plain. Thus, estuarine sediments serve as "leaky sinks" of trace metals to the continental margin.

A second pathway by which metals reach the continental shelf is through deposition of airborne particles. This mechanism has been implicated in explaining the elevated concentrations of such elements as Cu, Zn, Pb and Ni that may occur in the sea surface microlayer[25)], although bubble flotation also contributes to metal enrichment at the surface. Eventually, gravitational settling removes these particles to subsurface waters.

To delineate the behavior of as many metal ions, including pollutants, as possible, we select metals which are representative of the three biochemical classes of metal ions[26)]. These classes are characterized as oxygen-seeking, nitrogen/sulfur-seeking or of intermediate bonding preference. The first group is comprised principally of the alkali and alkaline earth metals, including two of considerable biochemical interest: Ba and Be. Because of its limited solubility in marine systems, Ba is generally considered to be of little toxicological interest. However, it does compete with Ca for selected anions and may interfere with metabolic processes in the cell involving Ca[26)]. Moreover, current research into the bioaccumulation of drilling mud components by bivalve molluscs has shown that these organisms accumulate Ba from suspensions of solid $BaSO_4$[27)].

Among the most strongly sulfur-seeking ions are those of silver, mercury and copper. Silver has become a model toxicant; its analytical chemistry is quite straightforward (it is one of the most sensitive elements in plasma emission measurements), and its background concentrations are very low. Because of the affinity of silver and other sulfur-seeking compounds for the –SH and –S–S– groups that are vital to protein integrity and enzyme function, these elements are notably toxic.

Most of the first row transition elements, and Cd, exhibit intermediate bonding preferences, sometimes occurring in chelation matrices containing both oxygen and nitrogen or sulfur ligands. Because Cd has been the subject of considerable toxicological study in marine systems, $Cd^{++}$ will be used as an example of a "borderline" metal ion.

Metal ions in the aqueous phase of sea water are potentially assimilable by living organisms; i.e., they are bioavailable. Metals absorbed on settling particles such as hydrous manganese and iron oxides, clay minerals and detrital materials, may have lost much but not all of their bioavailability. Availabilities to benthic fauna and to the larvae of these fauna depend on a number of ill-defined factors, including the particular metal involved and its physiological importance to the organism; the physical state and the availability of that phase to the specific organism; the concentration of the element in the environment and the presence of other elements that may inhibit or enhance its uptake; the morphology of the organism, its life history, its condition and age.

## IV. Toxicity and Sublethal Effects of Selected Metal Ions

To characterize the effects of metal ions on marine organisms, a reference point is needed. Metal toxicity studies with larvae reveal many effects of metal ions in short-term experiments, and provide this reference point. The bioregulation and detoxification of metal ions is then considered in detail by focusing on two pollutant ions, barium and chromium.

Established techniques for rearing bivalves in the laboratory have made it possible to provide embryos and larvae for study[28, 29]. These studies use an artificial seawater for stability, with the metal introduced as inorganic salts. They have defined $LD_{50}$'s (doses lethal to 50% of the population), for Hg, Ag, Zn, Ni, and Pb, with the results applicable to the synthetic medium used. MacInnes[30], using a response-surface analysis technique, was able to extend the studies to demonstrate the interaction effects of two or three metals in combination. In agreement with Calabrese[28, 29], the studies showed mercury was the most toxic to embryos of oysters, followed by copper and zinc, and, in general, the interaction effects of both nitrate and chloride salts were similar.

Still in question is the combined effects of certain metals in life stages to ascertain whether metals in combination act synergistically or antagonistically. These possible interactions are important because they can illustrate the ways in which toxic elements can impair metabolic processes that depend upon essential elements. Some of the interactions involve competition for ligands; others are noncompetitive. The occurrence of an interaction probably has implications for the metabolism of elements other than those primarily involved[31].

Finally, a distinction must be made between developing embryos and fully formed larvae, since the tolerance may be different. If the growth of a fully developed larva is retarded, it could prolong the pelagic life stage and promote loss through predation, disease and dispersion.

## V. Effects of Pollutant Metal Ions on Bivalves

### 1. *Bioregulation and Detoxification*

A factor contributing to the uncertainties of predicting the bioavailability of trace metals is the ability of some organisms to regulate effectively metal assimilation. Bioregulation appears to be most effective for those metals which, in low concentrations, are essential nutrient elements (e.g., copper and zinc). The mechanisms through which marine fauna regulate some metals and detoxify others have been the subject of considerable study. There appear to be two pathways in operation in some organisms:

(1) the complexation of metals by organic chelators (e.g. metallothioneins) rendering them chemically less reactive and so less biologically active, and

(2) the encapsulation, isolation and elimination of metal salts as discrete particles[32].

Metallothioneins, found in a wide variety of vertebrates, invertebrates and a few algae, are low molecular weight (6,000–10,000) proteins that are synthesized in response to sublethal exposure to various heavy metals. The majority of work on invertebrates has been done with mussels (*Mytilus edulis*) and oysters (*Crassostrea virginica* and *Ostrea edulis*). *M. edulis* cadmium-binding proteins appear to be "classical" metallothioneins. Preliminary analyses[33, 34] indicate that the proteins have a high proportion of cysteine and a typical methallothionein UV absorption. Purified *Mytilus edulis* metallothionein isolated from both the digestive gland[35] and kidney[36] has the following properties similar to mammalian metallothioneins: high cysteine content, higher absorbance at 250 nm than at 280 nm, and the ability to sequester cadmium, copper and zinc. In contrast, the cadmium-binding proteins in *Crassostrea virginica* are not true metallothioneins[37]. The low proportion of cysteine (8%) and the predominance of aspartic and glutamic amino acids suggest that the metalbinding capacity in the oyster is associated with carboxyl groups[38].

Aside from these apparent chelation mechanisms, many studies have shown that a variety of heavy metal pollutants are sequestered within deposits referred to as "granules", "concretions", or "membrane-bound vesicles" in various tissues of bivalve molluscs[5,39,40]. The concretions are predominantly composed of calcium, complexed with phosphate, but contain high concentrations of such heavy metals as Cu, Pb, Zn, Fe, Au, V and Hg. It has been proposed that metal storage in these concretions is a general detoxification mechanism[41]. Concretion formation, however, is not understood. It is suggested that concretions are synthesized, either totally or in part, via a lysosomal system within the epithelial cells of the kidney[36]. This model, however, does not explain how heavy metals are transported to the kidney for subsequent incorporation into the granules. Amoebocytes have been implicated[41], but it is not known whether the metals are complexed with calcium phosphate prior to or subsequent to this transport.

In some species, both mechanisms of detoxification (i.e. complexation by organic chelators and concretion formation) are present and apparently are interrelated. In *Mytilus edulis*, for example, cadmium is initially associated with high molecular weight proteins[42,43], but with time is transferred to low molecular weight metallothioneins[43]. It has been postulated[36] that the metallothioneins may be present within the concretion vesicles (whether or not in association with the actual granules is unknown), and is released when the membranes rupture during homogenization.

## 2. Experimental Studies

Two metallic components of drilling muds, barium and chromium, have been studied for their impact on bivale molluscs[44]. Generally, chromium bioaccumulation is limited to forms soluble in seawater, whereas barium, present as insoluble $BaSO_4$, is assimilated with a resultant elevated level in selected organs of the molluscs[45,46].

In molluscs the kidney was the only organ with measurable amounts of chromium. The level rose in molluscs selected for experimentation and exposed to chromium, but the levels were variable between organisms and different populations. This trend may reflect seasonal differences such as those observed by Bryan[47], who suggested that variations in the concentrations of several elements in *Chlamys opercularis* and *Pecten maximumus* were due to changes in the availability of food.

Barium, however, is taken up not only by the kidneys of molluscs, but also is measurable in the adductor muscles and digestive gland.

### 3. *Conclusions*

The results of sequential leaching experiments and the bioaccumulation studies reveal a selective assimilation of certain chemical forms of some trace metals by molluscs. There seems to be a relationship between seawater-soluble chromium levels and accumulation of the element in the kidneys of exposed bivalves. Particulate forms of chromium, while present in the digestive glands, do not appear to have been assimilated by the organisms, even though much of the total chromium was dissolved by a relatively mild extractant –25% acetic acid. Thus, an estimation of the bioavailability of particulate metals based on their solubility in dilute acetic acid can be erroneously high, at least for herbivorous filter feeders.

Particulate barium sulfate, however, is taken up by all three species. It is not clear whether the particles of barium sulfate were actually solubilized by, for example, enzymes that control calcium metabolism, before the barium was processed by the kidney. Since small increases in the barium content of the large adductor muscle of *P. magellanicus* accompanied large increases of barium in the kidney, it may be that the organism effectively shunted the element from digestive gland to kidney. This shunting has been studied in the isolation and detoxification of copper by "green-sick" oysters[48)]. The case for a shunting mechanism would be more compelling if other particulate metals, such as chromium, were similarly enriched in kidney tissues. That chromium levels in kidneys did not follow the total chromium loading to which the organisms were exposed puts this mechanism in doubt. Moreover, other particulate metals, namely iron and aluminum, were not found at elevated levels in the kidneys of test organisms.

## VI. Accumulation of Vanadium by Tunicates

Approximately two thousand species of tunicates are to be found in the world's oceans, many of them beautiful in form and color. They are small animals, the largest specimens seldom exceeding fifteen cm in length (Fig. 1). The name "tunicate" is derived from the tough, leathery outer covering that surrounds and supports the soft inner body. Tunicates are considered chordata, although primitive ones, because the larvae possess several structures, such as the notochord, which appear to be analogous to a chordate structure; in this case, a spinal cord. The common name "sea squirt" stems from the observation that intertidal species, when exposed above the water line, can contract, forcing sea water out of their bodies.

Tunicates feed on the microorganisms and organic detritus suspended in oceanic waters. Several families of these animals perform the remarkable feat of assimilating vanadium against an almost ten million-fold concentration gradient. In sea water, vanadium is present as the vanadate ion, principally $H_2VO_4^-$, at a uniform concentration of about $5 \times 10^{-8}$M[49)]. In vanadium-containing blood cells, called vanadocytes, the concentration of vanadium, usually present as V(III) or V(IV), can be as high as 1 M[50)]. How does this extraordinary accumulation take place? How do the internal and external

**Fig. 1.** *Ciona intestinalis.* This specimen was collected from a holding tank in the New England Aquarium. It is most likely that the larval stage passed through the sea water intake and filtering system, then attached and developed into an adult. A specimen collected directly from unfiltered Boston Harbor waters would be covered with debris

environments of tunicates affect this process? Before answering these questions, we briefly review some aspects of vanadium biochemistry and function, and tunicate biology.

## *1. Vanadium Biochemistry*

Only the features of vanadium biochemistry relevant to understanding the accumulation mechanism in vanadocytes are presented here. More extensive coverage of this subject may be found in two recent reviews[51, 52].

The presence of vanadium in plant and animal tissues had been recognized for some time[2], but relatively little was known about its biochemistry. As mentioned above, vanadium is spectacularly concentrated in vanadocytes. Tunicates appear to be unique in this respect, since vanadium occurs only as a trace element in other animals. However, in mammals it is an essential nutrient[53]; it is estimated that approximately two mg per day is the human requirement[54]. In these living systems, vanadium is present in an aerated, aqueous environment. Let us see what form we would expect for the element in this environment.

Four oxidation states of vanadium are fairly stable in aqueous solution[15]. At $[H^+]$ = 1 M the ions associated with these oxidation states ranging from (II) to (V) are $V^{2+}$, $V^{3+}$, $VO^{2+}$ and $VO_2^+$. Vanadous ion, $V^{2+}$, slowly reduces water to hydrogen gas. It is of no biological significance, and will not be considered further. Vanadocytes store vanadium chiefly as a complex of the vanadic ion, $V^{3+}$. The uncomplexed $V^{3+}$ ion hydrolyzes to $VOH^{2+}$ at pH2.2, then dimerizes and precipitates at higher pH. This behavior implies that the interior of the vanadocyte is either at an unusually low pH for a living cell, or the vanadic ion is complexed. As shown below, evidence favors complex formation rather than low intracellular pH.

The paramagnetic vanadyl ion, $VO^{2+}$, is a less exotic ion than $V^{3+}$ to contemplate in a living cell. In fact, owing to the convenience of its unpaired electron, vanadyl chemistry has been well-characterized over a wide range of pH values, and it is used extensively as a probe of biochemical structure and function[51]. It is, however, easily air oxidized to vanadium(V)[55], the species produced depending strongly on conditions. In sea water, for example, the anionic vanadate species, $H_2VO_4^-$ and $HVO_4^{2-}$, are ultimately encountered. Vanadate has a strong tendency to condense and form polyvanadates[15]; however, at the nanomolar concentration level in sea water, only monomers are present in appreciable amounts.

All three oxidation states of vanadium found in living systems form stable complexes with a wide variety of mono- and multidentate ligands. Ions of these three oxidation states form complexes rapidly; they are therefore labile. As mentioned on page 142, bimolecular complex formation rate constants for monomeric vanadate species are approximately $10^4$ $M^{-1}s^{-1}$. Water exchange rate constants are $280s^{-1}$ for $V^{3+}$, and $500s^{-1}$ for $VO^{2+}$ at 25 °C[52].

From a chemical point of view, the setting for the encounter between tunicate and vanadium at the sea water interface is fairly simple to visualize. In the oxygenated, mildly alkaline, highly dilute conditions of the external sea water environment vanadium is present as monomeric vanadate anion. It is there predominantly as the free ion; the low concentrations of organic chelators in sea water favor complex dissociation rather than formation (Cf. p. 141)[8].

Conditions in the internal tunicate environment can be, and proably are quite different. Organic acids with carboxyl, amine and/or hydroxy groups are present at concentrations sufficiently high to shift metal-ligand equilibria towards complex. Moreover, cells present environments of controlled electrochemical potential, determined by the presence or absence of redox active cell constituents such as $O_2$, ascorbic acid, glutathione, etc. The three vanadium oxidation states of interest interconvert rapidly[52], so that accumulated vanadium should exist as a complex, in one or more oxidation states. As an example of an application of these inorganic principles to the physiological role of vanadium, consider for a moment the element's biochemistry in mammalian muscle tissue.

Approximately six years ago vanadate was identified as a contaminant in commercially prepared (Sigma) ATP[56], that inhibits the sodium- and potassium-stimulated adenosine triphosphatase enzyme (Na, K-ATPase)[57], and vanadium ceased being a curiosity. Further investigation disclosed that the inhibition of the enzyme by vanadate could be reversed by a reducing agent of suitable potential, such as norepinephrine – a derivative of o-catechol[58]. From these reports it became obvious that the chemistry of vanadium allows it to play two roles in biology; it can act as a phosphate analog, and as a redox reagent. However, the low ($\mu$M) concentration of vanadate in mammalian tissues, combined with the relatively higher (mM) concentrations of reducing species such as glutathione, keeps most of the vanadium in the V(IV) state[4]. In this form, the element is tied up as a stable complex with hemoglobin[20], transferrin, or other proteins[51].

## 2. *Vanadium Function*

What then is the function of vanadium, so widely distributed among living things? Its role as a regulator of phosphatase enzymes was suggested by its ability to turn these enzymes off and on. However, with vanadium present chiefly as a complex of V(IV), a regulatory role does not seem likely. The function of endogenous vanadium in mammalian tissues remains unknown; perhaps its biological function will be revealed in studies on tunicates, where – in the blood cells – the concentration levels are so much higher than in mammalian cells and tissues. Tunicate studies have so far given some tantalizing hints as to what may be the solution to the vanadium function problem, but no definitive answers.

The obvious role for a metallic element in a blood cell, as an oxygen binder and transporter, has been ruled out experimentally. Let us briefly review the history of this proposed function. Henze, the physiologist who discovered vanadium in tunicates, speculated about the function of the element some twenty years after reporting his discovery, but did not advocate oxygen binding[59]. Later, less catious researchers, struck by the green color of some vanadocytes, and the discovery of nitrogen-bearing components of these cells, which were (incorrectly) identified as pyrroles, named the pigment responsible for the coloration "hemovanadin" on the basis of an analogy to hemoglobin[60]. It was suggested that oxygen binding was the function of this pigment, and several researchers attempted to determine the extent of this effect, but the experimental procedures were unclear, and the results were often in disagreement[61]. An *in vitro* demonstration of oxygen binding by a vanadium complex seemed to strengthen, indirectly, the validity of oxygen binding as the element's function[62]. Finally, relatively straightforward experiments were carried out to determine vanadocyte oxygen-binding ability.

The procedure used was to determine the concentration of dissolved oxygen ($O_2$) with an $O_2$-electrode[63]. A (linear) calibration curve was prepared of current (due to the response of the $O_2$-electrode) *vs.* oxygen concentration in sea water ranging from fully deaerated to saturation at an $O_2$ pressure of one atmosphere.

In the test for oxygen binding, $O_2$-electrode measurements were made by diluting samples of oxygenated sea water with equal-sized volumes of either deaerated sea water or deaerated vanadocyte suspensions. If vanadocytes could bind $O_2$, the concentration of dissolved $O_2$ should be lower in the sample diluted with vanadocytes, than in the one diluted by deaerated sea water. However, no difference in the post-dilution sea water oxygen concentration was found between the two; the results fell on the calibration curve. This finding has now been extended to tunicate blood cells rich in iron, rather than vanadium[64].

Another approach to function ascribes anti-feedant effects to vanadium[65]. Aside from weather, adult tunicates, which are sessile (stationary) animals, are primarily vulnerable to two types of ecological threats. They may succumb to predatory crustaceans and fish, or they may be fouled; i.e., covered over by growths of smaller organisms such as smaller tunicates, or algae. How do immobile tunicates survive these pressures? Is it possible that they use a chemical defense to thwart their enemies? If so, is vanadium their main line of defense?

Two types of evidence indicate that vanadium may function as a anti-feedant for unicates. First, there is evidence based on observation that species whose outer coverings (tunics) are rich in vanadocytes are shunned by predatory fish[65, 66]. These tunics are often acidic, which confers further protection from predation on an individual specimen exhibiting this characteristic. Since vanadium would be expected to hydrolyze in sea water, this acidity could result from release of vanadium due to rupturing blood cells. Second, pieces of fish deliberately dosed with appreciable amounts of vanadium, and offered as food pellets to crustaceans and fish, resulted in reduced food consumption by these predators[67].

However, the evidence linking the function of chemical defense to vanadium in ascidians is not entirely conclusive. The concentration of vanadium in tunicates is variable[67], and many species have vanadium levels which correspond to those dosages *not* observed to deter the food consumption of potential predators. Furthermore, success in survival does not correlate with high vanadium content, acidity, tunic toughness, or other such factors. Quite successful species such as *Ciona intestinalis*, with low vanadium content, neutral and unexceptionally tough tunic are as adaptable and widespread as many other species which possess high vanadium content, and acidic and exceptionally thick, tough tunics such as *Ascidia nigra*. However, the possibility that, in some species, vanadium provides a measure of chemical defense cannot be dismissed.

If, on the other hand, vanadium is present to carry out some biochemical function as Henze speculated[59], it may be that the process is tunic formation. There is circumstantial evidence through which vanadocytes have been associated with tunic formation. Microscopic studies have revealed that the ultimate fate of vanadocytes is to rupture inside the tunic, which is infiltrated with blood vessels. Emanating from the disintegrating cells are strands of tunic-like material, that become incorporated into, and are indistinguishable from the extant tunic[68]. Whole animal studies, in which pieces of glass were deliberately inserted into the tunic to form wounds, elicited a congregation of vanadocytes at the wound site, whereupon the wound was rapidly healed by formation of new tunic[69][1].

The association between tunic formation and vanadocytes was put in doubt when it was shown that radioglucose, incorporated into the tunic, was derived from underlying epithelial cells, and not from vanadocytes[70]. The tunic is a complex tissue, however, not composed of a single structure. It consists of an amorphous ground substance with considerable glycoprotein, fibers with mainly a cellulose-like composition, and a proteinaceous outer cuticle[71]. A recent study has shown that the glucose label enters the ground substance, and not the fibers[72]. Therefore, the possibility that vanadium is involved in tunic fiber formation cannot be eliminated.

1 This paper reported cellular responses in *Molgula manhattensis*, a species which contains ferrocytes, but no vanadocytes. We have repeated the experiments with *C. intestinalis*, a vanadocyte-containing species, and found responses similar to those described for *M. manhattensis*

At the present time, considerable insight into the biochemistry of vanadium in tunicates is associated with the mechanism of vanadium uptake. We shall concentrate on this topic, first presenting the biological context in which to view this remarkable feat of metal ion accumulation.

## 3. *Biological Factors in Vanadium Uptake by Tunicates*

Tunicates possess a circulatory system with a one-chambered heart, a nervous system with ganglia, a hermaphroditic reproductive system, and an endocrine system[73)]. They have two openings called siphons, which are protuberant in some species, less so in others[74)]. A ciliary pump located in the incurrent siphon draws in sea water, passing it through the digestive tract, and out to the external environment *via* the excurrent siphon.

Tunicates feed by filtering the sea water passing through their bodies. A net with rectangular pores of sizes from 0.2 to 0.5 $\mu$m in width and 0.5 to 2.2 $\mu$m in length secreted by the animal traps their food[75)]. Although this trapped material can, and probably does, constitute a particulate source of vanadium, tunicates are capable of accumulating vanadium from the aqueous phase of sea water[49)]. The final repository for this vanadium is mainly in vanadocytes (Table 1).

Experiments on whole tunicate blood are difficult to control because of the delicacy of vanadocytes. These cells cytolize (rupture) easily, changing the pH and color of the plasma, and inducing clumping and other changes to occur in the remaining intact cells. Careful experimentation, however, shows that, for most species studied, the blood plasma is colorless and iso-osmotic with sea water[77)]. Differences between plasma and sea water are found for pH and sulfate. Reliable pH measurements are in the range 6.9 to 7.4, and sulfate concentration is approximately 40% lower than sea water. However, the chloride concentration is elevated.

Many different types of blood cells have been identified in tunicates. Not all the functions of these cells are known[76)]. Some may be phagocytic, others may serve as repositories of waste products. Among the most numerous cells are the vanadocytes; spherical cells 8–12 $\mu$m in diameter that are packed with smaller (2–3 $\mu$m diameter) granule-filled vacuoles (Fig. 2). They acquired their name from osmium tetroxide staining studies that showed a reducing species, presumably vanadium, was present in these cells. Other tunicate species possess cells of a similar appearance which contain significant amounts of iron, but no vanadium; these cells have been termed ferrocytes.

The tunicate's discrimination among ions may take place immediately upon entry into the internal environment. Studies on the blood plasma[78)], and blood of certain Australian

**Table 1.** The Distribution of Vanadium and Iron in *A. nigra*[76)]

| Tissue | V Concentration | Fe Concentration |
|---|---|---|
| | mg/kg dry weight | |
| tunic | 320 ± 180 | 160 ± 40 |
| body | 2120 ± 560 | 230 ± 90 |
| plasma | 11 ± 4 | 0 |
| blood cells | 26760 ± 8900 | 270 ± 50 |

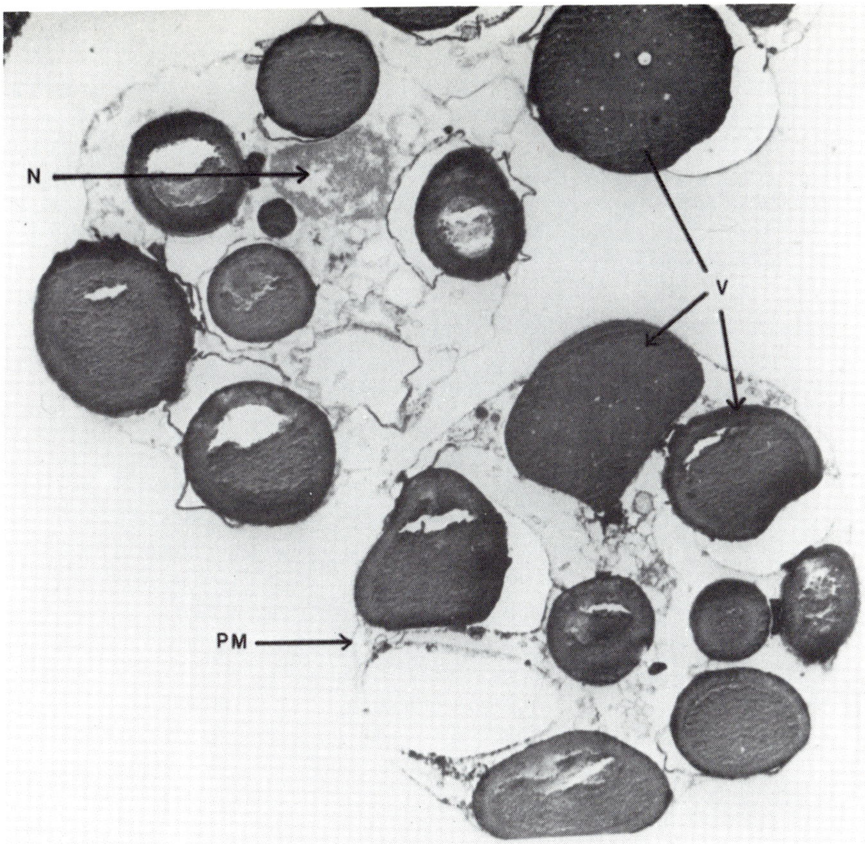


**Fig. 2.** Electron micrograph of vanadocytes from *Ascidia nigra*. Cells were fixed in 2.5% glutaraldehyde in 0.15 M HCl, followed by 1% osmium tetroxide, pH 7.6. The sections were stained with uranyl acetate: *N* = nucleus, *PM* = plasmalemma, *V* = vacuoles (very electron dense presumably because of reduction of osmium tetroxide by V(III).)

tunicates[79)], are indicative of such selective processes. No discrimination was found in the uptake of $^{59}Fe$, $^{54}Mn$, $^{51}Cr$, and $^{48}V$ at the sea water-plasma interface, but binding within the plasma was found to be selective, due to the presence of chelators with different affinities for the different metal ions. However, the ultimate rejection or assimilation decision resides in the blood cells.

## 4. *Vanadocytes*

Vanadocytes are nucleated, possess organelles, although few mitochondria[80)], and the conspicuous vacuoles. The cells of many species such as *A. nigra*[81)] and *A. ceratodes*[82)] are a bright yellow-green, due to the vacuolar coloration. These observations are drawn from microscopic examination of blood samples. It is more difficult to isolate vanadocytes from all other blood cells, and determine their chemical composition. However, since vanadocytes of mature specimens of many species comprise 50% or more of all blood cells[81, 82], 90+% preparations of vanadocytes can often be achieved. Determinations of vanadium content from native and enriched blood cell or vanadocyte samples indicate that the element is most concentrated in vanadocytes. Other types of blood cells contain vanadium, and it has been suggested that these cells are vanadocyte precursors[83].

By using non-invasive techniques, it is possible to determine the concentration and oxidation state of vanadium in living cells. In *A. ceratodes,* EXAFS (X-ray absorption fine structure) techniques showed that 90% of the vanadium is in the form of V(III), 10% is V(IV)[84]. In *A. nigra,* EPR (electron paramagnetic resonance) spectroscopy showed that 95% of the vanadium is in the V(III) form, and 5% is V(IV)[85].

Electron microscopic studies are not consistent with one another in regard to the site and distribution of vanadium in blood cells, however. In one study[86], cells stained with osmium tetroxide (which is presumably reduced by vanadium to osmium, which blackens the reaction site) were compared through examination by light and electron microscopy. In both techniques, the vacuoles were the darkest areas of the cells, with fine granules in the vacuolar interior accounting for most of the scattering. In another study, utilizing X-ray microanalysis of unstained preparations in scanning and transmission electron microscopy to detect vanadium[87], different results were obtained. Most of the vanadium detected was found on the vacuolar membranes of blood cells referred to as amebocytes, signet ring cells, and compartment cells, and scant amounts were found in vanadocytes. (In both studies the cell and vacuolar membranes appear to be simple lipid bilayers.) Clearly, more studies are required by these potentially accurate and informative methods to reconcile these differences. In this review we assume that vanadium is present in the vacuoles of the vanadocytes in appreciable quantities, and that it is also present in lesser, but detectable quantities in other cells.

Early studies on vanadoytes ascribed the green coloration and $OsO_4$ reducing capability to "hemovanadin"[60]. By implication, vanadium should have been bound to a macromolecule, probably a protein. However, dialysis, chromatographic, magnetic resonance and susceptibility, and spectrophotometric studies have shown that intracellular vanadium is not tightly bound to a macromolecule[50, 63, 84, 88]. The green color of the blood cell is not due to vanadium, but to "tunichrome", an organic molecule of relatively low molecular weight[76, 81, 89]. It has been suggested that tunichrome reduces vanadium(V) to vanadium(III)[81]. However, *in vitro* studies carried out thus far have only established that tunichrome is a rapidly reacting reductant capable of reducing V(V) to V(IV)[81].

The discovery of vanadium in tunicate blood cells by Henze was unusual enough[90, 92], but Henze found a second oddity. Cytolysis of these cells in distilled water produced acidic solutions, which extrapolated back to extraordinary acidities of 1–2 M in the intact cells. Subsequent findings that the vanadium was in the +3 oxidation state[88, 93], made these acidities seem plausible, since this oxidation state is unstable in aqueous solutions with $[H^+]$ above $10^{-3}$M. Such low intracellular acidities corresponding to 1–2 M sulfuric acid are not known in living cells, however, and the supposed non-existence of V(III) in any but very acidic media ignores the stabilizing effect of complex formation. It is now clear that cellular cytolysis results in the hydrolysis of tunichrome, V(III) and V(IV). Since tunichrome and *total* vanadium are roughly equal in concentration, both in the range 0.1–1 M, their hydrolytic reactions account for the increased $[H^+]$ of the extracellular medium following cell rupture.

Cytolysis wasn't the only method used to determine intracellular pH. In another approach, acid-base indicators such as methyl orange were diffused into living cell suspensions, and intracellular color changes corresponding to low pH were observed[94, 96]. All the indicators used also function as irreversible redox indicators, however. Therefore, it is most likely that these indicators measured cell potential, not pH.

To avoid these pitfalls two different techniques have recently been used to measure the intracellular pH in intact *A. nigra* blood cells[97]. One technique combines a measurement of the ratio of intracellular to extracellular concentration of amine with measurement of the extracellular pH to yield the intracellular pH. Another technique involves intracellular $^{31}$P-NMR (nuclear magnetic resonance) measurements. These methods led to a determination of pH7.39 $\pm$ 0.10 for unfractionated cells, and 7.2 for the vanadocyte fraction.

It therefore seems pertinent to ask how a cell with typical pH, chemical components, membranes, and organelles could accumulate a trace metal against an often more than 10 million-fold concentration gradient. The answer lies in the change of charge on vanadium associated with the reduction that accompanies the transport of the element from the external into the internal environment.

## 5. Dynamics of Vanadium Uptake by Vanadocytes

In the external environment of tunicates, most metal ions are cationic, few are anionic. Vanadium is one of the anions, as it is present predominantly as $H_2VO_4^-$ at the pH of sea water. Only chromate, among the metal-containing anions, is significant in sea water and it is present at a fiftieth the concentration of vanadium[8]. Other anions to consider are sulfate and phosphate, present as $SO_4^{2-}$ and $HPO_4^{2-}$ in sea water.

By using radioisotopes, the influx of vanadate into the blood cells of *A. nigra* has been studied[85]. The influx of phosphate, sulfate and dichromate, and the inhibitory effect of these oxoanions on vanadate influx have also been determined. The rate of vanadate influx was measured in the presence of metabolic inhibitors and inhibitors of anion transport. In this study, EPR spectroscopy was used to follow changes in the concentration of reduced vanadium within tunicate blood cells exposed to vanadate.

The influx of vanadate is a rapid ($t_{1/2}$ = 57 s at 0 °C), monophasic process[85], but the rate of vanadate influx approaches a saturation limit at high concentrations of exogenous vanadate. Influx experiments with added EDTA (ethylenediamminetetraacetic acid) were carried out to eliminate exchange or adhesion of vanadate to the cell surface to account for the reduction of radiovanadium in the extracellular medium. The amount of vanadium accumulated in a typical influx experiment was independent of added EDTA, and raised the intracellular concentration by 2.5 mM. Therefore, there is a *net* uptake of vanadium into the intracellular volume of the vanadocytes in these experiments.

Neither inhibitors of glycolysis, nor uncouplers of cellular energy-releasing processes had any significant effect on the rate of vanadate influx. Phosphate, which is readily taken up by the cells inhibited vanadate influx. Sulfate is not accumulated by the cells, and neither sulfate nor chromate appreciably inhibited vanadate influx at concentrations up to 10.9 mM and 200 $\mu$M, respectively. Most significantly, inhibitors of anion exchange across the human red cell membrane such as DIDS (4,4′ – diisothiocyanostilbene – 2,2′ – disulfonic acid) were also found to block vanadate transport into vanadocytes.

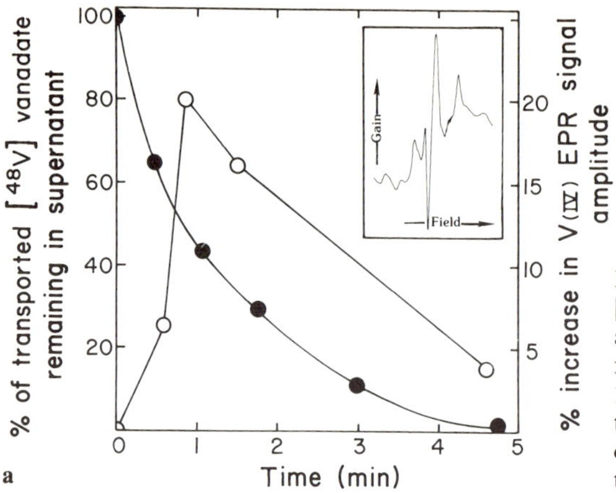


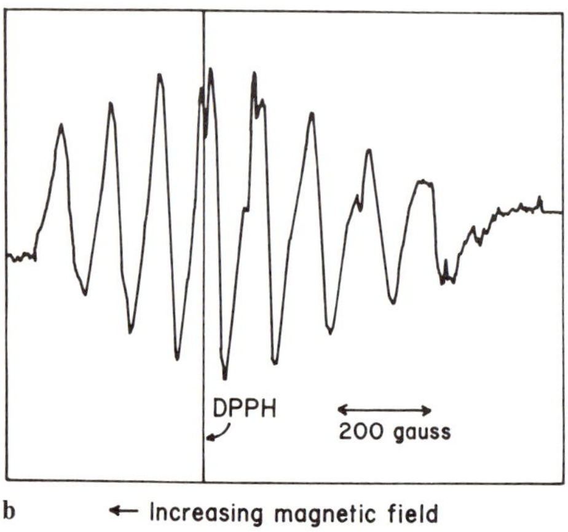


**Fig. 3. a** Increase in intracellular [V(IV)] during influx of V(V). (O–O); a 10% cell suspension was divided into five aliquots which were each made 1 mM in V(V) and frozen at timed intervals. A room temperature spectrum of the cell suspension gave the concentration of V(IV) at time = 0. Insert; a spectrum of the frozen cell suspension; power = 8.0 mW, frequency = 9.3 GHz, modulation amplitude = 10, time constant = 1 s, gain = 12.5. (●–●); a time course for [$^{48}$V] vanadate influx. **b** EPR spectrum of living blood cells. Washed cells were suspended in HEPES/NaCl buffer at a concentration of $6 \times 10^8$ cells/ml and loaded into a flat quartz cell. The spectrum was recorded at 9 MW microwave power, $\nu = 9550$ GHz, time constant = 1 s) modulator amplitude = 106 and gain = $1.25 \times 10^3$. Broadening of all peaks and doubling of the two central peaks indicates the presence of two slightly different V(IV) species

The natural level of vanadum(IV) in *A. nigra* vanadocytes is about 5% of the total intracellular vanadium. Unlike V(V) and V(III), which are EPR-silent, V(IV) is EPR-active. The concentration of vanadium in this oxidation state was monitored by EPR spectroscopy during the course of influx. It was observed (Fig. 3) that the EPR signal of cell suspension samples increases at first, and then decreases during the period of vanadium influx.

A hypothesis has been presented, linking accumulation to vanadate reduction[81)]. The influx data support this hypothesis. *Selective* accumulation of vanadium appears to result from two factors.

(1) Vanadate enters the cell through anionic channels. This process eliminates positively charged metal ions, or metal ion complexes like $FeCl^{2+}$ present in sea water.

(2) Vanadate is reduced to vanadium(III); as a cation this species is not transported through anionic channels, and is trapped inside the cell, where it accumulates (Fig. 4).

Accumulation appears to be driven by the rapid reduction of intracellular vanadate to V(IV). Since the influx of ($^{48}$V)-vanadate occurs as a single exponential step, and is paralleled by the appearance of V(IV), it follows that the rate-limiting step of accumula-

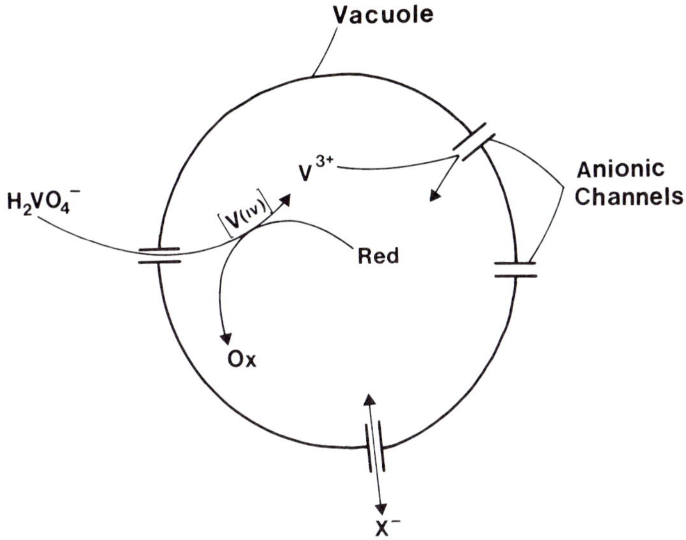


**Fig. 4.** Diagram of accumulation mechanism. Vanadium enters the vacuole within the vanadocyte as mono-negative $H_2VO_4^-$, although it may be possible for the dinegative anion, $HVO_4^{2-}$, to enter this channel as well. ($X^-$ stands for any negative ion such as $Cl^-$, $H_2PO_4^-$, etc., that may exchange across the membrane through the anionic channel). Reduction to $V^{3+}$ takes place in two steps, *via* a V(IV) intermediate. The resulting cations may be trapped as tightly bound complexes, or as free ions which the anionic channel will not accept for transport. The nature of the reducing species is unknown at present

tion is the entry of vanadate into the cell. The electron donor for the reduction of V(V) to V(IV) is unknown, but could be tunichrome. Another role for tunichrome might be to complex, and stabilize V(III). Attempts to isolate tunichrome would lead to vanadium-free preparations if the complex's stability constant were moderately, but not excessively high, as V(III) is so labile.

The assimilation of vanadium by tunicates into vanadocytes clearly involves two of the three accumulation processes mentioned earlier: membrane transport and intracellular reduction. Many questions remain, however, concerning the role of complex formation in the initial stages of transport through the environment/blood plasma barrier, and in the nature of the intracellular reduction of V(V) to V(III). These, and the questions of vanadium function and intracellular form should yield to future investigations.

*Acknowledgements.* The tunicate studies were supported by a research grant form the National Science Foundation (K. K. and G. C. M.), and were carried out with the collaboration of Drs. A. Dingley and I. Macara.

## VII. References


1. Kustin, K., McLeod, G. C.: Topics Curr. Chem. *69,* 1 (1977)
2. Meisch, H.-V., Bielig, H.-J.: Basic Res. Cardiol. *75,* 413 (1980)
3. Eigen, M.: Pure Appl. Chem. *20,* 93 (1969)
4. Langford, C. H., Gray, H. B.: Ligand Substitution Processes, New York, W. A. Benjamin, Inc. 1965
5. Ducommun, Y. et al: Inorg. Chem. *19,* 3696 (1980)
6. Swaddle, T. W.: Coord. Chem. Rev. *14,* 217 (1974)
7. Oser, B. L.: Hawk's Physiological Chemistry, p. 322, New York, McGraw-Hill Book Company 1965
8. Horne, R. A.: Marine Chemistry. New York, Wiley-Interscience 1969
9. Lehninger, A.: Biochemistry, p. 783, New York, Worth Publishers, Inc. 1975
10. Walshe, J. M.: Wilson's disease, a review. In: The Biochemistry of Copper (Peisach, J. et al., eds.). New York, Academic Press 1966, p. 475
11. Basolo, F., Pearson, R. G.: Mechanisms of Inorganic Reactions, p. 223, New York, John Wiley and Sons, Inc. 1967
12. Margerum, D. W. et al: Kinetics and Mechanisms of Complex Formation and Ligand Exchange. in: Coordination Chemistry. Vol. 2 (Marlell, A. E., ed.), Washington, D. C., Amer. Chem. Soc. 1978, p. 163
13. Püschel, R., Lassner, E.: Chelates and chelating agents in the analytical chemistry of molybdenum and tungsten. In: Chelates in Analytical Chemistry. Vol. 1 (Flaschka, A., Barnard, A. J., Jr., eds.). New York, Marcel Dekker 1967, p. 265
14. Ferguson, J. H., Kustin, K.: Inorg. Chem. *18*, 3349 (1979)
15. Cotton, F. A., Wilkinson, G.: Advanced Inorganic Chemistry, New York, Interscience 1980
16. Vold, R. R., Vold, R. L.: J. Magn, Reson. *19*, 365 (1975)
17. Gilbert, K., Kustin, K.: J. Am. Chem. Soc. *90*, 5502 (1976)
18. Funahashi, S. et al: Inorg. Chem. *20,* 1752 (1981)
19. LaGrange, P.: unpublished
20. Cantley, L. C., Jr. et al: Biochim. Biophys. Acta *629*, 95 (1980)
21. Cantley, L. C., Jr.: Curr. Topics Bioenergetics *11*, 201 (1981)
22. Knauf, P. A.: Curr. Topics Membranes Transport *12,* 249 (1979)
23. Turekian, K. K.: Geochim. Cosmochim. Acta *41*, 1139 (1977)
24. Olsen, C. R. et al.: Estuar, Coast. Mar. Sci. *10*, 119 (1980)
25. Reference 1., p. 4
26. Nieboer, E., Richardson, D. H. S.: Environ. Poll (Series B) *1*, 3 (1980)
27. Gilbert, T. R. et al.: In: Wastes in the Ocean, Vol 4, Energy Wastes in the Ocean, (Duedall, I. W., Kester, D. R., Park, P. K., and Ketchum, B. H.) Wiley-Interscience, NY, In Press, 1983
28. Calabrese, A. et al.: Mar. Biol. *18*, 162 (1973)
29. Calabrese, A., Nelson, D. A.: Bull. Environ. Contam. Toxicol. *11,* 92 (1974)
30. MacInnes, J. R. Mar. Environ. Res. *4*, 217 (1980)
31. Sandstead, H.: Interactions of Toxic Elements with Essential Elements. In: Micronutrient Interactions: Vitamins, Minerals and Hazardous Elements, Ann. N. Y. Acad. Sci. *355* (1980)
32. Briggs, LeB. R.: Unpublished, 1981
33. Noel-Lambot, F.: Experienta *32,* 324 (1976)
34. Talbot, T., Magee, R. J.: Arch. Environ. Contamin. Toxicol. *7*, 73 (1978)
35. George, S. G. et al.: Biochem. Biophys. Acta *580,* 225 (1979)
36. George, S. G., Pirie, B. J. S.: Biochem. Biophys. Acta *580*, 234 (1979)
37. Ridlington, J. W., Fowler, B. A.: Chem.-Biol. Interact. *25,* 127 (1979)
38. Engal, D. W., Fowler, B. A.: Copper and Cadmium-Induced Changes in the Metabolism and Structure of Molluscan Gill Tissue in Marine Pollution: Functional Responses (Vernberg, W. B., Calabrese, F. P., Thurberg, F. P., Vernberg, F. J., eds.), New York, Academic Press 1979
39. Rheinberger, R. et al.: The Kidney of the Quahog (*Mercenaria mercenaria*) as a Pollution Indicator. In: Animals as Monitors of Environmental Pollutants, Nat. Res. Council, Nat. Acad. Sci. 1979

40. Carmichael, N. G. et al.: Fish. Res. Bd. Can. *36*, 1149 (1979)
41. Coombs, T. L., George, S. G. Mechanisms of Immobilization and Detoxication of Metals in Marine Organisms. In: Physiology and Behavior of Marine Organisms. (McLusky, D. S., Berry, A. S., eds.), New York: Pergamon Press 1978
42. Scholz, N.: Helgol. Meeresf. *33*, 68 (1980)
43. Carpene, E., George, S. G.: Molec. Physiol. *1,* 23 (1981)
44. Gilbert, T. R. et al.: Availability of Trace Metals in Drilling Fluids to Marine Bivalve Molluscs, Northeast Shellfish Conf., Edgartown, MA., Sept. 1981
45. Knox, F.: Thesis, Massachusetts Inst. of Technology 1978
46. Chow, T. J., Goldberg, E. D.: Geochim. Cosmochim. Acta *20,* 191 (1960)
47. Bryan, G. W.: Mar. Bio. Ass. U. K. *73,* 145 (1973)
48. George, S. et al.: Mar. Biol. *45*, 147 (1978)
49. Kustin, K., et al.: J. Gen. Physiol. *65,* 315 (1975)
50. Carlson, R. M. K.: Proc. Nat. Acad. Sci. (USA) *72,* 2217 (1975)
51. Chasteen, N. D.: Structure and Bonding *53*, 105 (1982)
52. Kustin, K., Macara, I. G.: Comments on Inorganic Chemistry (1982), in Press
53. Mertz, W.: Science *213,* 1332 (1981)
54. Schwartz, K.: Fed. Proc. *33*, 1748 (1974)
55. Dean, G. A., Herringshaw, J. F.: Talanta *10,* 793 (1963)
56. Cantley, L. C., Jr. et al.: J. Biol. Chem. *253*, 7361 (1978)
57. Cantley, L. C., Jr. et al.: ibid. *252*, 7421 (1977)
58. Cantley, L. C., Jr. et al.: J. Am. Chem. Soc. *100*, 5210 (1978)
59. Henze, M.: Hoppe-Seyler's Z. Physiol. Chem. *213*, 125 (1932)
60. Webb, D. A.: J. Exp. Biol. *16,* 499 (1939)
61. Senozan, N. M.: J. Chem. Ed. *51,* 503 (1974)
62. Swinehart, J. H.: Chem. Commun. 1443 (1971)
63. Macara, I. G. et al.: Comp. Biochem. Physiol. *62A*, 821 (1979)
64. Agudelo, M. I. et al.: ibid. *72A*, 161 (1982)
65. Stoecker, D.: Mar. Ecol. Prog. Ser. *3*, 257 (1980)
66. Stoecker, D.: J. Exp. Mar. Biol. Ecol. *48*, 277 (1980)
67. Stoecker, D.: Ecol. *61*, 1327 (1980)
68. Endean, R.: Q. J. Microscop. Sci. *102*, 107 (1961)
69. Anderson, R.: Biol. Bull. *141*, 91 (1971)
70. Wardrop, A. B.: Protoplasma *70,* 73 (1970)
71. Patricolo, E., DeLeo, G.: Acta Zool. *60*, 259 (1979)
72. Robinson, W. E. et al.: J. Exptl. Biol. (1983), in Press
73. Van Name, W. G.: Bull. Amer. Mus. Nat. Hist. *84*, 1 (1945)
74. Abbott, D. P., Newberry, A. T.: Urochordates. In: Intertidal Invertebrates of California (Morris, R. H., Abbott, D. P., Haderlie, E. C., eds.), Stanford, CA, Stanford University Press 1980.
75. Flood, P. R., Fiala-Medioni, A.: Acta Zool. *62,* 53 (1981)
76. Macara, I. G. et al.: Comp. Biochem. Physiol, *63B*, 299 (1979)
77. Goodbody, I.: Adv. Mar. Biol. *12,* 1 (1974)
78. Hawkins, C. J. et al.: Biol. Bull. *159*, 656 (1980)
79. Hawkins, C. J. et al.: ibid. *159*, 699 (1980)
80. Deck, D. J. et al.: J. Morph. *120,* 267 (1966)
81. Macara, I. G. et al.: Biochem. J. *181,* 457 (1979)
82. Biggs, W. R., Swinehart, J. H.: Experientia *35,* 1047 (1979)
83. Kustin, K. et al: Biol. Bull. *150,* 426 (1976)
84. Tullius, T. D. et al.: J. Amer. Chem. Soc. *102,* 5670 (1980)
85. Dingley, A. L. et al.: Biochim. Biophys. Acta *649*, 493 (1981)
86. Gansler, H. et al.: Experientia *19,* 232 (1963)
87. Botte, L. et al.: ibid. *35,* 1228 (1979)
88. Boeri, B., Ehrenberg, A.: Arch. Biochem. Biophys. *50*, 404 (1954)
89. Gilbert, K. et al.: J. Cell Physiol. *93*, 309 (1977)
90. Henze, M.: Hoppe-Seyler's Z. Physiol. chem. *72*, 494 (1911)
91. Henze, M.: ibid. *79,* 215 (1912)
92. Henze, M.: ibid. *86,* 340 (1913)

93. Resaeva, L. T.: Fedn. Proc. Fedn. Am. Soc. Exp. Biol. Suppl. *24,* 836 (1964)
94. Hecht., S.: Am. J. Physiol. *45,* 157 (1918)
95. Endean, R.: Aust. J. Mar. Freshwater Res. *6*, 35 (1954)
96. Endean, R.: Q. J. Microscop. Sci. *100*, 177 (1960)
97. Dingley, A. L. et al.: Biochim. Biophys. Acta, *720*, 384 (1982)

## Author-Index Volumes 1–53